# Universities and Climate Action

# Universities and Climate Action

Tristan McCowan

UCLPRESS

First published in 2025 by
UCL Press
University College London
Gower Street
London WC1E 6BT

Available to download free: www.uclpress.co.uk



A CIP catalogue record for this book is available from The British Library.



ISBN: 978-1-80008-827-6 (Hbk)
ISBN: 978-1-80008-828-3 (Pbk)
ISBN: 978-1-80008-829-0 (PDF)
ISBN: 978-1-80008-830-6 (epub)
DOI: https://doi.org/10.14324/111.9781800088290

# Contents

*List of figures* vii
*List of tables* ix
*Acknowledgements* xi

1 Introduction 1

2 Climate change as a civilisational crisis 15

3 Knowledge and climate change 31

4 Modalities and pathways of impact 55

5 Global dynamics of higher education 81

6 Teaching climate change 103

7 Curriculum topography 131

8 Institutional embodiment 153

9 Engaging with society 175

10 The road ahead 199

*References* 211
*Index* 225

# List of figures


4.1 University modalities 59
4.2 Stages of impact of the university on climate change 64
4.3 Mitigation and adaptation pathways for university influence on climate change 67
5.1 Internationalising the education modality 96
5.2 Dynamics of impact through internationalisation 99
7.1 Curriculum topography in the university 139
9.1 Dynamics of university–society engagement 190

# List of tables


4.1 Key to mitigation and adaptation pathways for university influence on climate change 68
4.2 Positive impacts of universities on climate change 70
4.3 Negative impacts of universities on climate change 72
7.1 Topographical features in four curricular interventions 144

# Acknowledgements

Many of the ideas in this book were formed during the period of the Climate-U project, and so benefited from the numerous interactions with researchers in Brazil, Fiji, India, Indonesia, Kenya, Morocco, South Africa, Tanzania and the UK, and in some cases from direct experiences in those countries. The dialogue, immersion, inspiration and friendship were all crucial to the book, and I am grateful to my many guides in the different terrains. While this book does not report on the empirical findings of Climate-U, it does present the theoretical framework I developed as part of the project and therefore benefited from funding from the UK's Economic and Social Research Council via the Global Challenges Research Fund. Some of the chapters draw on previously published material: Chapter 4 on *The Impact of Universities on Climate Change: A theoretical framework* (Transforming Universities for a Changing Climate, Working Paper Series No. 1, 2020); Chapter 5 on 'Internationalisation and climate impacts of higher education: Towards an analytical framework' (*Journal of Studies in International Education*, 2023); Chapter 6 on 'The climate crisis as a driver for pedagogical renewal in higher education' (*Teaching in Higher Education*, 2023); and Chapter 7 on *Climate Change in Higher Education: A curriculum topography approach* (Transforming Universities for a Changing Climate, Working Paper Series No. 6, 2021). My thanks go to: Rebecca Schendel, Kate Roach, Keri Facer, Simon Marginson and Charlotte Nussey for their comments on drafts of the above papers; to David Heymann for graphic design of the figures appearing in this book; and to the anonymous UCL Press reviewer for a range of helpful observations on the manuscript. I am grateful to Pat Gordon-Smith at UCL Press for her support and guidance throughout the publication process. Finally, my thanks go to my family, and particularly its newest member, Freya, whose first few years of life have accompanied the writing of this book.

1

# Introduction

At 7.16pm on Saturday, 12 December 2015, the green gavel of Laurent Fabius, the French foreign minister, finally came down to signal the approval of the Paris Agreement,[1] to the cheers and tears of the exhausted delegates. After the disappointments of the Copenhagen conference six years earlier, the agreement of 196 parties[2] to cut emissions and maintain temperatures at no more than 1.5°C above pre-industrial levels was heralded as a triumph of diplomacy and collective will. Yet it remains a fragile accord. Fulfilment of the required targets is only a voluntary commitment of each nation-state. There is no enforcing power and no meaningful penalties and incentives, other than the collective benefits of averting a climate catastrophe – some way in the future, and beyond the timeframe of electoral considerations and accountability of politicians. The task still remains then to find a way to bring about the substantial changes that are necessary – in individual behaviour, in culture and custom, business and trade, policy, science and technology, in all the varied sectors of society and parts of the world.

In the year the Paris Agreement was being signed, 217 million students were enrolled in tertiary education institutions around the world, soon to step out into their lives as graduates, professionals, leaders and citizens (World Bank 2024). That figure is increasing year on year, representing a proportion of nearly 42 per cent of the global cohort (UIS 2023a). Their experience at university has profoundly influenced their outlook on life, their skill sets, knowledge base, social networks and expectations. Individually and collectively, university has equipped them with the apparatus either to continue destroying the environment for immediate material gain, or to move towards and beyond the requirements of the Paris Agreement, in ensuring a sustainable human society and planet. If we are to avoid both a climate catastrophe and the imposition of authoritarian rule to force a reduction in emissions, then we need to harness the generative powers of universities – powers that have been, and to some extent continue to be, yoked to the plough of fossil fuels instead.

This is a book about the complex and contradictory roles that universities play in climate change. Through their diverse functions of providing general education for increasing proportions of the population, delivering professional training, conducting basic research and technological innovation, engaging with the public and offering services to government, private sector and civil society organisations, they give us the tools to both destroy and save ourselves. Ultimately, we must choose, but we must also understand what that choice involves. What are the pathways through which higher education influences climate change? Can we enhance its positive impact, and if so how?

## Throwing down the gauntlet to higher education

Universities have been paying increasing attention to sustainability and climate change in recent years. A number of national and international alliances have been created to promote environmentally friendly actions in universities, such as the International Universities Climate Alliance, the Sustainable Development Solutions Network and the Association of University Leaders for a Sustainable Future. University leaders have attempted to map and align their work with the Sustainable Development Goals (SDGs), to promote research addressing local and global sustainability challenges, to engage with communities, and to promote understanding of climate change through the curriculum. Efforts have also been made to reduce the carbon footprint of campuses and to regenerate local areas, and many universities have set targets for carbon neutrality and net zero carbon;[3] some such as the University of Plymouth and the London School of Economics and Political Science in the UK have already been certified as carbon neutral. The generally high levels of environmental awareness and concern among young people have meant that the green credentials of universities have also made them more attractive to prospective students.

Nevertheless, there has not yet been a genuine transformation of the sector, and many of the above initiatives – while sincere and well intentioned – have remained piecemeal and on occasion tokenistic. The regime of marketisation, which characterises the global system as a whole (though with substantial variations across contexts), has meant that student recruitment, capturing of research funds and other income-generating activities have trumped other concerns. In some instances (for example, when green credentials are an effective marketing ploy), these interests can dovetail, but there are often conflicts and trade-offs, with public good benefits losing out. Most scholarship on the issue (for example, Facer 2020; Leal Filho 2010; Rapley et al. 2014; Binagwaho et al. 2022) has shown that competing priorities and barriers to engagement with society have weakened universities' potential to be protagonists in climate action.

As much as those working in universities may hold dear the humanistic, emancipatory aims of the institution, it cannot be denied that the staggering growth of the sector in recent decades has been due to quite different reasons. The possession of a university degree has become the ultimate passport into salaried employment, serving either as a professional certificate in itself (such as for medical staff, architects or engineers) or a generic qualification signalling aptitude and diligence for prospective employers. This tacit pact between higher education and the labour market has brought prominence and revenue to the sector, but at a cost. The investments made by governments (in those contexts in which public support still exists), or by students and their families, inevitably bring expectations of the delivery of career success – expectations that can never be fulfilled for all. These pressures – combined with the more brutal commercialisation of the for-profit higher education sector – serve to stifle both the traditional academic pursuits of the institution (in medieval or Humboldtian vein) as well as the engaged, public good mission that can support sustainability and climate action.

The obstacles faced by higher education institutions (HEIs) in bringing a transformation (in themselves and in the outside society) in line with the demands of sustainability are, therefore, considerable. Yet there is a more fundamental problem, which is that we lack a clear understanding of the ways universities might have an impact on climate change in the first place. Universities are primarily institutions for the generation and propagation of ideas. This process takes place through: the transfer, debate and interpretation of existing knowledge (most obviously through the 'teaching' function, and present since the medieval universities); the storing, preserving and curating of that knowledge (a function that is becoming increasingly obsolete in the internet age); and the generation of new knowledge (through the 'research' function, and associated primarily with the Humboldtian university from the nineteenth century).

Given the materiality of climate change – the increases in average temperatures brought about by the build-up of greenhouse gases in the atmosphere – what relevance might this dance of ethereal ideas have? One answer is that it was ideas that put us in this predicament in the first place. The roots of climate change are in the development of our understanding of the natural world and its biological, chemical and physical dynamics over the last half-millennium, along with the emergence of a set of values holding that this scientific understanding should be applied to the improvement of the human condition – with the human conceived as being separate from the rest of nature, entitled to exploit the natural world for its benefit and dependent primarily on material wealth and comfort for its well-being. Addressing climate change, therefore, also relies on ideas – the challenging of the values and technologies that made possible the Industrial Revolution, and the putting in place of a new set that can support the regeneration of human communities and the natural environment.

Yet universities have also developed stronger material ties with society. Changes in the university from the nineteenth century onwards (which materialised in the land-grant universities in the United States of America (USA), the engaged public universities in Latin America following the Córdoba reforms and the developmental universities in Africa; see Chapter 9) have led it into a closer relationship with production, incorporating new areas of study relating to agriculture and industry, and more extensive applied research and innovation (Bernasconi 2007; Carpentier 2019; McDowell 2003; Perkin 2007). Universities, therefore, have also become sites for development of products and processes that can directly exacerbate or mitigate greenhouse gas emissions.

Furthermore, universities are increasingly taking on a role of influencing those outside the institution, and not in any obvious relationship to it – through secondments and consultancies to governments, private sector and civil society organisations, through public engagement in mainstream and social media, and through various forms of partnerships and projects. These new relationships have meant that universities do not only release ideas and their human embodiments into the broader society but are actively involved in the application and use of those ideas.

Universities today, therefore, maintain a range of interlocking relationships with climate change, some positive and some negative (from the perspective of ensuring humanity's continuing existence), which involve our collective understanding about climate itself, the development of the knowledge, skills and values of individuals for addressing the challenge, and the production of technologies and direct interventions at different scales. Furthermore, universities are themselves communities with their own environmental and social dynamics and impacts. The complexity of these roles, relationships and influences makes it challenging to determine what the impact of higher education is, and unclear what interventions or changes would be advisable at a particular point in time. This book aims to contribute to this task. By presenting a theoretical framework of the university in relation to climate change, it provides a basis for understanding the dynamics and – if not a prescription for action – at least a platform on which action can be deliberated and decided on.

Overall, this book argues that universities have a crucial role in humanity's quest to address the climate crisis. The transformative power they have – both in relation to the individuals who pass through them as students and collectively for our knowledge and understanding of the world – has the potential to turn us from the destructive path we are currently following and towards a sustainable way of living with each other and with the planet. Yet it is potential, not inevitability. To fulfil it, universities must go through their own transformation.

## What we know about higher education, sustainability and climate change

Literature on climate action in higher education has been growing in recent years. A systematic review produced by the Climate-U[4] project showed a year-on-year increase in the number of Web of Science articles published on the topic, from just one in 2003[5] to 24 in 2019 (Nussey et al. 2023).[6] The articles were distributed between different functions of the university, with 65 on curriculum and pedagogy, 3 on knowledge production, 36 on community engagement, 10 on public debate and 37 on campus operations – representing a varied if uneven spread of academic attention. Many of these studies document initiatives of specific universities, often those in which the authors themselves are located. While the review only focused on active responses of universities to the climate crisis, there is also a small body of literature assessing the negative impacts of universities from their emissions – for example, through their campuses (Royal Anniversary Trust 2023), international student travel (Shields 2019) and academic conferences (Bjørkdahl et al. 2022).

In addition to these empirical studies, there are some normative treatments, advocating for particular positions or approaches. Facer (2020) provides a comprehensive assessment of the transformation needed in universities, covering: the reconfiguring of operations to reduce emissions; the strengthening of institutions' civic role; reshaping knowledge structures to acknowledge interdisciplinary complexity; and refocusing the educational mission to support living together with the planet. Stein and colleagues (Stein 2019; Stein et al. 2023) provide a decolonial critique of conventional sustainability efforts in universities, arguing that the climate crisis is rooted in colonisation, and that transformation will only be possible through recognition of ongoing coloniality and its uprooting. Lotz-Sisitka, Macintyre and colleagues (Lotz-Sisitka et al. 2015; Macintyre et al. 2018) advocate for a 'transgressive' approach to pedagogy, to disrupt dominant ideas of 'resilience' that fail to address the roots of unsustainability. There are also some theoretical accounts focusing on particular dimensions of the issue, such as Molthan-Hill and colleagues (2019) with their categorisation of curricular structures, and Frediani and Nussey (2021) on university–community engagement for climate action.

There are also a small number of book-length treatments of the topic. *Universities on Fire: Higher education in the climate crisis* by Bryan Alexander (2023) analyses the range of future scenarios facing universities, and the devastating impact that environmental changes will have on their campuses, finances and modes of operation. Some edited collections of academic works have been published, gathering together for the most part case studies of initiatives in different institutions and contexts. For example, Reimers (2021) addresses the transformation of school curricula and non-formal education through constructive engagement with universities, and brings together a range of international perspectives from Guatemala, Haiti, Pakistan and the USA, while Leal Filho (2010) focuses on teaching and learning within the university, providing 21 case studies from around the world.

These works are embedded in a broader literature on higher education and sustainability – involving climate but also a range of other environmental and social questions, framed particularly in relation to the SDGs. Universities have found in the SDGs a convenient framework for mapping their existing provision in order to demonstrate the public benefit generated from their activities, and a normative scheme for aligning future work. An academic journal has been established to focus on the topic (*International Journal of Sustainability in Higher Education*), publishing studies on campus sustainability, the curriculum, assessments and rankings, and whole-institution approaches, among other themes.

There is also a body of literature on climate change education at the school level. As will be explored in greater detail in the chapters that follow, this literature has provided a range of evidence on effective and ineffective practices relating to learning about climate change (for example, Bangay and Blum 2010; Læssøe et al. 2009; Rousell and Cutter-Mackenzie-Knowles 2020), many of which are also relevant for the teaching of adults in higher education. Stuart Tannock's (2021) book *Educating for Radical Social Transformation in the Climate Crisis* explores how currents in egalitarian, antiracist, anticolonial and feminist education can be engaged to bring the much-needed transformation for climate justice.

While the body of literature is growing and providing us with a firmer basis for understanding and action, there are still some significant gaps. As seen above, certain aspects of university activity remain understudied – for example, the role of universities in influencing public opinion, and the setting of research agendas. In terms of geographical focus and authorship, there is – as in all areas of publication – a bias towards high-income Global North countries and institutions, despite the active and impactful work carried out by many universities in the Global South. Many of the studies outlined above focus only on one dimension of the university; there is also a need, therefore, to conceptualise and research the different functions of the university as they work together, and understand the relationships between them. It is this last point that will be the primary focus of this book.

## Understanding the role of universities in climate action

In summary, then, why a book on climate change and higher education? In part, the answer is that attention needs to be paid to every facet of society and aspect of our lives if we are to address the 'wicked' problem. But there is something particular, even unique, about universities as institutions and the part they must play in addressing the climate crisis. First, universities have already played a prominent role in the story of climate change. They have been an active party in the establishment and globalisation of fossil-fuel-based industrialisation, in terms of scientific and technological breakthroughs, and the propagation of the skills and knowledge needed to drive that industrialisation forward (Bell 2021). In recent years, they have partly redeemed themselves, through generating the evidence base and analysis needed to show the link between greenhouse gas emissions and climate change, the negative impacts of rising temperatures, and the ideas and technology needed to counter them. Nevertheless, universities are still a major part of the motor of unsustainable, exploitative capitalism, and disentangling them is key to dismantling that system.

Second, the climate crisis will only be solved with imagination and creativity. The university is still the primary knowledge institution in contemporary society (in generating, disseminating and validating knowledge), despite seeing some aspects of that primacy undermined in recent years through developments in information technology, and other social and educational currents (McCowan 2019; Santos 2004). The complexity of the climate system and its multiple intersections with complex societal systems and with human psychology and behaviour mean that extensive, in-depth and ongoing enquiry, research and innovation are required to understand and address it.

While the actions and decisions of a few powerful people can do much to either sink or save humanity, ensuring a sustainable planet ultimately requires the collaboration of all. For this reason, universal formal education plays a vital role in any response to the climate crisis, as acknowledged in all declarations and agreements stretching back to the United Nations Framework Convention on Climate Change (UNFCCC). Basic education for children is undoubtedly central in this role, being in many cases universal and compulsory, and reaching humans at a particularly absorptive, malleable and formative stage of their development. Yet higher education has taken on increasing relevance in this respect, on account of the stupendous expansion of the sector in recent decades, moving from 10 per cent of the global youth population in 1970, to 20 per cent in 2000 and over 40 per cent in 2022 (UIS 2023a). While still an elite experience in some contexts, higher education is increasingly becoming part of the general educational trajectory of young people, and an essential part of their civic development and initiation into society. As such, it is crucial that climate literacy and empowerment for personal and collective change for sustainability is incorporated into the university experience.

Finally, further to these particularities of higher education, it is also important to look at universities as we might do any other space in society: as an institution, a set of buildings and a human community. Universities are variously villages, towns or cities with their own greenhouse gas emissions, areas of urbanisation and wildlife, and friendships, creativity and injustices. Chapter 8 on institutional embodiment looks at these dynamics, the impacts that they have, and the ways in which they can be shaped to become part of the solution rather than the problem.

This book deals frequently with abstract ideas, and it might reasonably be asked why a theoretical or academic book on this topic is needed. Climate change is real, universities need to cut emissions and teach their graduates to be environmentally friendly, so should we not just get on and do it? Is part of the problem, perhaps, that academics are thinking and talking themselves into knots while the egg timer runs down? These questions are understandable, and the lack of action in most segments of society has been frustrating. However, it is dangerous to slip into an opposition between theory and practice. We need both of them, and rather than detracting from each other they can be mutually reinforcing. When dealing with systems as complex as those of climate and higher education, it is particularly important that we have a clear, profound and nuanced understanding of the interplay of the different elements in order to respond to their constantly shifting and emergent nature. Paulo Freire (1970) expressed this point in terms of the indivisibility of reflection and action, stating in *Pedagogy of the Oppressed*:

> When a word is deprived of its dimension of action, reflection automatically suffers as well; and the word is changed into idle chatter, into verbalism, into an alienated and alienating 'blah.' It becomes an empty word, one which cannot denounce the world, for denunciation is impossible without a commitment to transform, and there is no transformation without action. On the other hand, if action is emphasized exclusively, to the detriment of reflection, the word is converted into *activism*. The latter – action for action's sake – negates the true praxis and makes dialogue impossible. (Freire 1970, 87–8)

Moreover, theory is not an option, but imbues everything we do. As argued previously (McCowan 2015a), while formal theories may exist in a rarefied academic space, there is another sense in which theory is part of all action and thought. Whether witting or not, human interactions are based on abstractions from the particular – explaining and predicting phenomena and providing norms to guide behaviour – so it is not a question of whether we engage with theory but of whether we make already existing theory explicit and knowable.

Having said that, this book will not be purely theoretical but will engage with a range of contexts and institutions in order to provide examples and illustrations and show the messiness and dilemmas of real situations. While not reporting on specific research findings, the book draws on my experience working with universities across the world – in particular in Latin America, Africa and the UK, and to a lesser extent

Asia, the Pacific, continental Europe and North America – most recently as coordinator of the international research project Climate-U. The cases discussed in this book are illustrative, and all contexts are unique, so the ideas presented here need to be debated and localised in relation to the particularities of different higher education systems and institutions.

A theme that has run through my previous books on education and its links to citizenship, human rights and the SDGs (McCowan 2009; 2013; 2019) is the holding in creative tension of the analytical and the normative. Understanding how the world works has inevitable implications for the ways we should act within it, and, correspondingly, deciding on the best course of action depends on a clear vision of the state of things. This book aims primarily to provide an analytical framework – to chart the different ways in which universities have an impact on climate change – but inevitably implications of a normative nature arise, concerning what universities should be doing, how they should be changing, and what those working and studying within them and beyond should be doing about it.

## Scope and structure of the book

A note is needed here on the usage of the terms 'higher education' and 'university' in this book. HEIs take a range of different forms: some are comprehensive and have the full range of functions, including research, postgraduate provision and community engagement programmes, while others specialise in particular elements, with only undergraduate teaching or specific disciplinary professional areas. The interest of this book is to a large extent in understanding the interactions of the diverse functions, so primarily focuses on comprehensive institutions, though in full awareness that not all fit this mould. The term 'university' is used generically to refer to higher education institutions, although in some countries the term may be reserved for specific designations (for example, only research-intensive institutions).

Equally, clarification is needed over terms used to refer to the climate. The term 'global warming' became popular from the 1980s, but the more neutral 'climate change' has largely taken its place in recent years (the year 2014 was the crossover point when there started to be more Google searches for 'climate change' than 'global warming'). 'Climate change' came into the ascendancy for a variety of reasons: its veneer of technical neutrality made it seem more scientific than the more casual sounding 'global warming', and the changes in question may involve cooling of temperatures in specific places and at specific times as well as warming. It is possible that sceptics and deniers may have promoted the term 'climate change', amenable as it is to interpretation as a non-human process of periodic cooling and warming, although some denial books prefer to use 'global warming' (for example, Booker 2009).

In fact, many climate activists today reject both of these terms. Advocacy groups prefer the expressions 'climate crisis' or 'climate emergency' in order to express the urgency of the situation, and to distance themselves from the idea that climate change might be a natural and inevitable phenomenon, occurring over the long run in cycles that humankind has adapted itself to in the past. 'Climate justice' is preferred by those highlighting the relational disparities in fault and repercussions both within and between countries, the plight of the 'most affected people and areas' (MAPA), and the historical colonial and continuing neocolonial dimensions. In this book, along with 'climate action' to designate responses made, the vaguer term 'climate change' predominates, to ensure as broad an engagement as possible with the different ways in which the phenomenon is and can be conceptualised and operationalised in higher education. However, as an author I endorse the position of the activists critical of the term, and 'climate crisis' and 'climate justice' are also used on a number of occasions during the book, particularly in the more normative sections.

The book is broad in its geographical reference. As stated above, my own work has spanned various continents, and those experiences have informed the theoretical ideas presented, as well as providing material for examples and cases. Normatively, the work is also based on the idea that universities of quality are needed in every context – as part of the fabric of societies, but also specifically in order to address climate change. Nevertheless, there is not one global experience of higher education, and while some parts of the framework may have universal application, the argument presented here must be understood on the basis of contextual difference, and the particular histories of colonisation and exploitation, the unique cultural make-up of each place, and the current political and economic circumstances.

As explored in a previous book (McCowan 2019), the form of higher education originating in medieval Europe that we call 'university' – the dominant model today – is just one of a range of historical models of higher education, and indeed just one of the possible models that could characterise higher education in the future. In some cases, the Western-style institutions have ridden roughshod over existing local forms of higher education, and in others the closing of the higher education imagination has prevented the emergence of new forms (including in so-called Western contexts). We have, therefore, a surprising homogeneity of higher education globally, despite superficial differences between universities across different regions. This lack of diversity is a problem, but not one that should call us to abandon the institution altogether. There is room for transformation of existing institutions – as seen through current moves for decolonisation – as well as for the creation of new forms at the edges of the mainstream system.

A final point about the scope of this book is that the main focus is on the impact of universities on climate change, and not vice versa. The question of how climate change will affect the university is an important area of study too, and is addressed in works such as Alexander (2023) and Phelan and Lumb (2021): future challenges will include extreme weather and other direct environmental impacts on campuses, as well as the financial hit affecting their budgets. While the cyclical nature of the relationship between university and society/natural environment is acknowledged in the theoretical framework presented in Chapter 4, most of the attention in this book is on the role universities play in addressing the climate crisis.

The 10 chapters of this book can be grouped into three sections. The first section provides the backdrop to the topic, with this introduction in Chapter 1, followed by an outline of the factual basis of the climate crisis and its roots in contemporary civilisation in Chapter 2, and an exploration of its epistemic dimensions, contestations and denial in Chapter 3. The second section puts forward the theoretical framework underpinning the book: Chapter 4 provides a model of the distinct functions of the university and their impact on climate change, while Chapter 5 opens the lens to the global level, looking at how institutions work together through their interlocking systems. The third and longest section draws out the implications of the climate crisis for the different dimensions of the institution. Chapters 6 and 7 look at teaching/learning and the curriculum respectively, while Chapters 8 and 9 focus on the internal workings of the institution and its engagement with external communities. Finally, the conclusion in Chapter 10 draws together the diverse threads of the book, highlighting the task facing higher education institutions in the twenty-first century.

## Notes

1 Adopted at the United Nations Climate Change Conference (COP21).
2 195 countries and the European Union.

3 Carbon neutrality allows offsetting carbon emissions through carbon reduction projects elsewhere; net zero carbon is more demanding and involves either no emissions or direct removal of carbon from the atmosphere.

4 Climate-U (Transforming Universities for a Changing Climate) was a research project that ran from 2020 to 2024 funded by the UK's Global Challenges Research Fund, and involving 16 HEIs in Brazil, Fiji, India, Indonesia, Kenya, Tanzania and the UK. It continues as a network with an expanding number of members: https://www.climate-uni.com/.

5 None were recorded before 2003.

6 Note: The review only covered meta-analyses of university research on climate change, not climate science itself, which of course would constitute a much larger body of literature.

# 2
# Climate change as a civilisational crisis

Human catastrophes come in different guises. Some, like the bubonic plague or Spanish flu, ravage populations causing intense misery and fatalities, and then – after a period of time – are gone. Others, such as famine and drought, recur frequently throughout history, but can often be prevented or softened by good planning and management. Natural events, such as earthquakes and volcanic eruptions, are hard to mitigate but generally affect only geographically local populations. Catastrophes for humanity can either reduce quality of life or wipe life out, or both. They differ in relation to the proportion of the population they affect, the timescale across which they operate, the level of forewarning and the ease with which they can be resolved.

Climate change is a crisis with a set of characteristics that make it both particularly devastating and particularly hard to respond to. Ultimately, the rising temperatures will make Earth uninhabitable for human beings. The direct impact of the heat, along with a series of knock-on impacts – the melting of the icecaps, rising sea levels, extreme weather, biodiversity loss and disruption, drought, threats to agriculture, ocean acidification, spread of diseases – will increase human suffering and jeopardise survival. While most of the negative impacts are currently falling on the most vulnerable populations, and will continue to do so, ultimately all human beings will be affected.

Climate change (in its current stage at least) is less obviously visible than a plague or a volcanic eruption, and so easier to dismiss – in the context of psychological difficulties in dealing with long-term risks (Marshall 2014). The cause–effect link is more indirect, and therefore easier to deny. The build-up of greenhouse gases in the atmosphere has been intensifying for 300 years, yet despite scientific evidence emerging in the 1800s, it was only in the second half of the twentieth century that the damaging impacts became widely recognised (Bell 2021). Even at the time of writing in 2024, when the idea is globally accepted (UNDP and University of Oxford 2021), the message does not appear to have penetrated very deeply, given the alarming lack of action in practice.

But that is not the only challenge with climate change. Even once the penny has dropped and everyone is on board, making the necessary changes in practice is not straightforward. It may appear a simple problem if we think purely in terms of reducing the quantity of greenhouse gases in the atmosphere: the options available to us are ceasing to emit the gases in the first place or removing them from the atmosphere. Yet stopping emitting greenhouse gases involves a number of changes that have a direct bearing on humanity's current interests, and are defended by the same logic that would lead us to take action against climate change in the first place (namely, humanity's survival and well-being). Many of the technological and other innovations that could be developed – both to reduce emissions and capture them, and to address in other ways the environmental impacts of greenhouse gases – have unpredictable outcomes within a complex planetary system. To cap it all, there is a cumulative impact, meaning that the later any action is taken, the harder the problem is to resolve. In light of these factors, it is comprehensible – although not forgivable – that as a species we are falling short in our response.

This book is not a work of environmental science, and its contributions are directed at the educational sphere. But it is important at the outset to lay bare the ecological assumptions underpinning it. This book understands climate change as a civilisational crisis. This phrase might be understood as indicating that our very civilisation is at risk from the impacts of climate change – and this is undoubtedly true. Yet it is also meant in the sense that the crisis is *within* our civilisation – understood as the complex of modernity in its structural, physical and ideological aspects – and to a large extent caused by it. The roots of the crisis are deep in our way of life – in the structures of our societies; in the economic, political and cultural spheres; and in our conceptions of ourselves and our sets of values. 'Our' is clearly a loaded term in this context, and many would rightly challenge both this level of planetary unity and equal share of responsibility across all world regions. While the 'West' is a highly problematic concept – and is constituted through influences that come from outside early modern Europe when Western hegemony solidified – it is the Western way of life (with its combination of internal assumptions about the relationship between humans and the non-human world, and its external manifestations of industrial society) that is most to blame for our current predicament.

How we understand the roots of climate change is significant and affects how we deal with it. The implication of climate change being a civilisational crisis is that technical fixes will not be enough. A number of solutions have been proposed, and many more are still under construction, which involve geo-engineering of various forms – in order to capture more carbon from the air or prevent heat from entering the atmosphere. Many more technological innovations focus on ensuring more efficient energy use or providing alternatives to fossil fuels (Gates 2021). These technological contributions are vital and are certainly part of the solution – but they will never be the whole solution. Our current civilisational model, oriented around maximising consumption and accumulation, will always tend towards planetary destruction, even if periodically mitigated by technological fixes.

This chapter provides a brief summary of the environmental aspects of climate change, the movements of global temperatures, their causes and likely future trajectories – but for more comprehensive accounts readers should refer elsewhere, to the Intergovernmental Panel on Climate Change (IPCC) and many other climate scientists and activists. The task of this chapter is to draw out the implications of this scenario – with all its uncertainties and contestations – for the work of the university. In doing so, the account will highlight five key characteristics of climate change: (i) its *anthropogenic* nature, that is to say, its causes in human society, in economic, political and cultural structures we have created, and in our own being; (ii) its *complexity*, in the technical sense of being part of a complex rather than a linear system, and thereby resistant to simple interventions with predictable outcomes; (iii) its *contested* nature, in bringing disagreement among people in terms of the causes, current state of play and possible solutions; (iv) its *global* reach, being a challenge that goes beyond national boundaries; and (v) its *time-bound* nature, urgent and sensitive to timescale. These are not the only five characteristics of climate change. Nevertheless, they are the ones that have particular relevance for the work of the university and, as argued in the final section, both make the university essential to climate action and require a fundamental change in its modus operandi.

## The basics

The phrase 'climate change', in its current usage, is a shorthand term that refers to those changes in the Earth's climate attributable to human beings in the contemporary era, involving an overall increase in temperatures and knock-on environmental effects. Temperature rises are caused primarily by the emission of greenhouse gases from the burning of fossil fuels, most important among these being carbon dioxide, but also methane, nitrous oxide, chlorofluorocarbons and water vapour, which trap heat in the Earth's atmosphere. This warming is problematic for humanity for a variety of reasons, including rising sea levels, disruption of agriculture, extreme weather and loss of biodiversity (Anderson, K. 2012; Berners-Lee 2019; Klein 2014, 2019; Thunberg 2024).

Average temperatures on Earth have fluctuated considerably in the course of human history, moving between ice ages of various lengths (the last ending around 10,000 years ago) and warmer periods – a fact that has provided some ammunition for climate change deniers. Yet a particular kind of change has been observed since the seventeenth century, with the growth of capitalism and the exploitation of natural resources, brought to the global level through colonialism and an intensification of international trade. Geologists have started to call this period the Anthropocene,[1] succeeding the Holocene, characterised for the first time by shifts in the geological record that are primarily due to human causes (Lewis and Maslin 2018).

The Industrial Revolution in the eighteenth century led to the growing use of coal, then oil, then fossil gas[2] to power factories, vehicles and homes, and produce electricity. While there was immediate awareness of the local environmental (and social) damage caused by industrialisation, it was not until the nineteenth century that awareness started to be raised of a possible planetary impact on climate. In 1856, the US scientist Eunice Foote presented a paper at the American Association for the Advancement of Science showing that carbon dioxide in a glass container heated up more quickly than air when placed in sunlight (Bell 2021). Irish physicist John Tyndall obtained similar results a few years later, showing the infrared radiation absorption of different gases in the atmosphere. At the end of the century, Svante Arrhenius asserted that the burning of coal would lead to catastrophic global warming. Yet these ideas had little uptake or impact in mainstream society for many more decades. Ironically, it was the oil industry that generated much of the early research on the dangerous impact of fossil fuels on the atmosphere, although for obvious reasons they did little to publicise it. It was only in the 1980s when broader awareness in society started to develop, through events such as NASA scientist James Hansen's widely reported statement to a congressional hearing that human-caused global warming was now beyond doubt. From the 1990s, environmental issues started to move up the agenda, particularly through the 1992

Earth Summit in Rio de Janeiro, as public understanding and concern about climate change grew steadily. At the time of writing, while there is still significant contestation over responses to climate change – as explored in Chapter 3 – there is broad scientific consensus on the causes and impacts of the phenomenon, and very wide global public awareness (UNDP and University of Oxford 2021).

Climate change is not the only environmental problem facing us at the current time. The problems of chemical fertilisers and pesticides brought to prominence in Rachel Carson's (1962) *Silent Spring* have not left us; large cities, such as Delhi, Beijing and Lahore, have intolerable levels of air pollution; and efforts to limit the use of plastic bags and straws have done little to stem the build-up of plastic waste in the oceans. These and other environmental and social challenges should not be ignored in an exclusive focus on greenhouse gases. Yet for the most part there are strong synergies (tragic ones) between climate change and these broader concerns: water scarcity, biodiversity loss, desertification and deforestation are problems in their own right, but are also exacerbated by, and exacerbate, climate change. The only viable approach, therefore (and one that adds to its complexity), is in addressing these problems simultaneously, maintaining awareness of possible trade-offs. Tree-planting schemes may support mitigation through absorbing carbon but, if they rely on monoculture forests of pine and eucalyptus, may have other undesirable environmental impacts and lead to biodiversity loss.

Climate change is frequently described as a 'wicked problem'. Unlike conventional 'tame' problems, wicked problems cannot be solved by a technical 'engineering' approach, as they are complex, resist clear definition, are grounded in value perspectives and have innumerable potential solutions that cannot be pre-tested (Head and Alford 2015; Rittel and Webber 1973). The causes of climate change are multiple, its impacts are gradual and not easily attributable, and interventions in one area may bring unexpected changes in another area and cancel out any positive effect. Climate change has even been designated a 'super wicked' problem, with four aspects, namely:

> [T]ime is running out; those who cause the problem also seek to provide a solution; the central authority needed to address them is weak or non-existent; and irrational discounting occurs that pushes responses into the future. (Levin et al. 2012: 124)

When it comes to responses to the wicked problem of climate change, a commonly made distinction is that between mitigation and adaptation:

> Mitigation objectives address the causes of climate change, whereas adaptation objectives address the impacts of climate change through an adjustment in natural or human systems in response to the actual or expected climatic stimuli or their effects, which moderates harm or exploits beneficial opportunities. (Alves et al. 2020: 193)

The kinds of actions involved in mitigation and adaptation in higher education are likely to be different. Mitigation involves lessening the direct contribution of universities to climate change (through greenhouse gas emissions, investments in fossil fuel companies and so on), developing research and innovation in relation to fuel efficiency, carbon capture and similar mitigating activities, and changing the mindsets of students so as to encourage climate-friendly actions in their later lives. Adaptation, which is connected to ideas of preparedness and resilience (Holloway and Fortune 2018; Kitagawa 2017; Preston et al. 2015), will involve application of knowledge to address required changes in lifestyles, agriculture, housing, healthcare and so forth, both in relation to capacity-building and awareness-raising but also the generation of new ideas and technologies. Adaptation should not, however, imply surrender in the face of insurmountable odds, or a politically disempowering acceptance of the status quo: it is not an alternative to mitigation, but stands alongside it. Even as we transform our societies to aim for a more sustainable future, we will still need to adapt to changes in the climate already under way. While mitigation and adaptation are distinct, some actions will simultaneously address both: tree planting, for example, can serve both to absorb carbon (mitigation) and to address the impacts of climate change through providing shade and protecting water storage in soil (adaptation).

Theorists, such as Keri Facer, have added a third element: regeneration, or 'producing social systems or land use systems that are able to create positive benefits for restoring biodiversity, healthy ecosystems and viable communities' (Facer 2020: 15). Mitigation and adaptation alone fall short because they assume that heading off the worst of the calamity and keeping something like the status quo is enough. We need something more than that – regenerating our environment, putting in place the conditions for rich biodiversity, reforestation, replenishing of aquifers and ocean ecosystems. Regeneration relates not only to the natural environment but also to communities. Just as the climate crisis has been caused by a breakdown in human relations with the non-human world, so a sustainable planet depends on recovering harmonious relationships, not only with nature, but also between human beings – addressing inequalities, exploitation and exclusion. The presence of regeneration is important not only in providing conceptual and practical completeness to the task facing us in climate action, but also because it brings an element of hope and positivity. Instead of the doom and gloom narrative of a desperate scramble against melting ice caps, rising sea levels and global conflict, it turns the focus to the creation of a world that we would want to live in, one that may be better than the one we have now.

All universities have some responsibilities for adaptation, mitigation and regeneration, both in relation to themselves as institutions with their own communities, and in assisting communities in the society outside. Yet there are geopolitical differences here: universities located in wealthy neighbourhoods and in high-income countries may have a greater responsibility in relation to mitigation, as their local/national communities are likely to be disproportionately contributing to greenhouse gas emissions. Conversely, those universities located in lower-income areas and countries may have to work harder in relation to adaptation and regeneration, as their populations are likely to be disproportionately affected by the adverse impacts of climate change, and have fewer resources with which to combat them. Having said this, the pressures of economic growth at all costs are also strong in lower-income countries, and climate change impacts will be evident everywhere, so mitigation, adaptation and regeneration are necessary in all contexts.

The sections that follow will outline five key characteristics of climate change, before drawing out their implications for the role of the university.

## Key characteristics of climate change

### Anthropogenic

One key aspect of climate change in relation to the role of universities is its anthropogenic nature. While the general dynamics of climate are the result of the interaction of a number of factors, the vast majority of climate scientists attribute the recent increases in average temperatures to human causes (Oreskes 2004). The greenhouse gas emissions that are the direct cause are themselves rooted in the growth of industry, fuelled by consumerism and the capitalist system. While the rapid global population growth of the twentieth century inevitably placed a strain on the Earth's resources, it is in fact the excessive consumption of a small percentage of that population that is predominantly to blame for climate change.

Yet the roots of the climate crisis run deeper even than the structures of our societies, to touch our inner selves, our beliefs, values and understandings of the world. While there is still a diversity of worldviews between different cultural groups globally, the dominant one is an ontological/epistemological/axiological bundle that sees the human being as separate from the rest of the natural world, privileging human needs (and even wants), and thereby justifying human exploitation of the natural world. The logic of the superiority of certain beings has also facilitated the systematic exploitation of human communities for centuries, but while slavery and serfdom are at least rhetorically (if not always in practice) rejected in today's world, there is no such disapprobation when it comes to the non-human world.

If the causes of climate change are in the human worldview and societal structures, then so are the solutions. Transformation at all of these levels and in all of these facets of human life is, therefore, needed. A multisectoral approach becomes necessary, simultaneously acting in all spheres of society, and transforming our institutions – both in the sense of formal organisations and social norms. As might be expected, this type of change is extremely difficult to achieve. Individual and collective behaviours, attitudes, values and worldviews are ingrained, and meaningful changes require intense work on the self. The incentive structures of the capitalist economy make most pro-environmental and pro-social activities somewhat 'irrational'. The current nation-state system promotes international competition of a military or economic nature and makes any attempt at coordinated action (such as the agreements at the annual Conferences of Parties, or COPs) fiendishly difficult.

As will be drawn out further below, the anthropogenic nature of climate change has profound implications for the university. Understanding the causes and impacts of climate change involves not only the full range of life sciences, physical sciences, engineering and technology disciplines, but also economics, social sciences, arts and humanities (Leal Filho et al. 2018). It also requires attention to – and possibly integration of – the diverse functions of the institution: conducting science, educating the young, informing the public and working directly with communities.

## Complex

Climate change is also characterised by complexity. Climate is a 'complex' – rather than a 'complicated' – system, as it is not just made up of a large number of elements interacting in intricate ways but has inherent unpredictability and no clear chain of cause and effect (Tikly 2019). It is a non-linear system, with positive and negative feedback loops. Ironically, the positive loops are the ones we need to worry about: for example, increasing temperatures warm the ocean, but warmer oceans are less able to absorb carbon dioxide, thereby leading to greater build-up of the greenhouse gas and higher temperatures, in a self-reinforcing cycle. Complex systems are also characterised by *autopoiesis* – literally, self-production or self-organisation, referring to their ability to maintain themselves through adapting to changing circumstances – and by *emergence*, the appearance of new characteristics and dynamics as a result of the interaction of constituent parts.

There are many factors at play in the climate system – including the sun, but also the Earth's atmosphere, the Earth itself, the oceans, ice, plant and animal life – and aspects of their interaction that are hard to identify and predict. Denial of anthropogenic global warming has been aided by the fact that the Earth's climate has changed repeatedly in past millennia on account of natural causes. Most contemporary scientists working on questions of climate are convinced that temperatures are being increased through human activity, yet there is still some uncertainty as to the speed of those changes, the impacts and the interventions that might mitigate them.

Solutions for the crisis are hard to identify, not only because they require multisectoral action, but because we cannot always be sure whether interventions in one area will not bring unexpected outcomes in another (Leal Filho et al. 2018). There is widespread concern about geo-engineering interventions, such as the use of aerosols to block the sun's rays, on account of the potential knock-on impacts on other aspects of the environment – for example, monsoon rains (Klein 2014). Berners-Lee (2019) discusses the disheartening phenomenon of rebound effects, through which efficiency gains in energy usage have led not to decreased usage of fuels, but to increased energy consumption. These elements of complexity are central to the 'wicked' or 'super wicked' designation of the climate crisis.

University researchers and teachers, like anyone else, frequently think, write and act in linear ways, and much ink has been spilt on the challenges of silo working in higher education, of disciplinary divides supported by the structures of taught courses, research funding and career incentives. Shifts towards addressing complexity, to holistic thought and action, and transdisciplinary enquiry are challenging for individuals and institutions. Yet the potential of universities is immense in this regard, given their core purpose for deepening human understanding, as well as the resources they offer for sustained enquiry and the different perspectives they bring to bear.

## Contested

On account of the profound implications of climate change for all aspects of our lives, there have been various forms of resistance and contestation. Climate change is contested in three ways: in its facts, in the strategies to address it and in the normative questions it mobilises. It is well known that, despite the vast majority of scientists asserting the existence of anthropogenic global warming, there are contrary voices (for example, Booker 2009; Lomborg 2007; Morano 2018; Koonin 2021). Some of these argue that temperatures are not rising at all, others that they are rising but due to natural not human-made reasons, and others that anthropogenic global warming is a reality, but that the impacts will not be as severe as made out by the likes of Al Gore and his *Inconvenient Truth*. These disputes have been intensified by the epistemic polarisation, the fuelling of distrust of experts and questioning of scientific knowledge in the 'post-truth' era.

Even among those who do recognise the reality of global warming, there is significant contestation over what should be done about it. Some see technology as the answer, resting their faith in future technological advances in the areas of carbon capture or geo-engineering. Others (including most environmental organisations) see that the reduction in emissions is essential, and most governments recognise the need for a movement towards renewable energy sources. Some (for example, Orr 1994) go much further than this position and argue for the need for a veritable paradigm shift for human society, a move away from consumerism and the forging of a new relationship with the natural world, of harmony and non-exploitation. These different responses are closely linked to an understanding of the problem as, on the one hand, primarily scientific, or on the other, being more strongly rooted in society, in politics, economics and culture – but they are also predicated on diverging values of a moral, political and even aesthetic nature (Marshall 2014).

While outright denial may appear to have become a marginal position (with the fossil fuel lobby moving instead to a strategy of *delay*), it lurks under the surface, and can reappear at any moment. An example here is the reaction to the Just Stop Oil protests in the UK in 2022, when deniers quickly regrouped and trumpeted their messages on social media, emboldened by the general popular disgruntlement at disruption in the movement of traffic and interruption of sporting events.

Contestation is a fact of life in higher education. Scholars have always disagreed, and those disagreements, while at times childish and egotistical, have spurred us on to extraordinary discoveries and creations. Open debate in the context of different value positions and contested facts is also a key part of teaching and learning. While suffering constraints on academic freedom in practice across the world, either from repressive states outlawing the discussion of certain topics as a result of taboo or social norms, or from academic dogmatism, universities still represent important spaces for respectful and constructive contestation. These forms of contestation will be discussed in greater detail in Chapter 3, which focuses on the epistemic dimension of the crisis.

## Global

The global dimension of climate change is perhaps the most obvious. While countries and continents have particular climatic conditions, they are locked into a planetary system from which it is impossible to delink. In the fictional account of future climate crisis in Kim Stanley Robinson's (2021) *The Ministry for the Future*, India takes unilateral action following a deadly heatwave and implements its own emergency geo-engineering response, blocking out some of the sun's rays. Inevitably, the impacts go beyond India's borders – interventions of this kind can never be limited to the national sphere. This simple fact has highly complex ramifications. On the one hand, it significantly raises the stakes on national responsibility, since decisions taken by leaders (and possibly citizens and consumers) in one country affect not only their population but the rest of the world. The current situation of climate injustice is largely the result of a lack of responsibility of this type on the part of the wealthiest communities and countries: if they could plead ignorance until some decades ago, they certainly cannot now. On the other hand, it raises the potential that one country or institution might find a solution that will save the whole world.

Yet is that really possible? This kind of assumption certainly underpins initiatives to find all-encompassing technological solutions of the kind promoted by Richard Branson with his US$25 million Virgin

Earth Challenge prize for the best carbon capture design.[3] Faith in this kind of solution is also tempting for business leaders and others in the Global North, as it simultaneously supports their self-belief in their privileged agency to determine the world's destiny, and also absolves them from making any fundamental changes to their practices, comforts and profits.

Nevertheless, this kind of outcome is highly unlikely. Even if the challenge were solely that of mitigation, it would still need a remarkable scientific breakthrough, and very rapid product development, to move us to a fossil-free energy regime across the globe in a short period of time. Moreover, the recent history of technology has taught us that breakthroughs are rarely shared equitably among all people, regardless of wealth and power. However, mitigation is not our only challenge. Adaptation to new climatic conditions is already necessary for many communities around the world, and will become increasingly urgent in the coming decades. Regeneration of the natural environment and of human communities' relationships with each other and with the non-human world is also vital if we are to build a future that is sustainable – not only in terms of averting the climate catastrophe but also for the full range of other environmental challenges facing us. Adaptation and regeneration – by definition – need to be carried out in situ, because they vary so much from place to place. And because they are intrinsically human transformations, they need to be centred in the agency of each community, rather than brought about by an external country, institution or plan. So the solutions to climate change may be variegated in terms of different forms of influence but inevitably they will involve all communities in the globe.

There are corresponding ramifications for the university. Many highly ranked HEIs in the Global North pride themselves on addressing grand challenges and finding global solutions to the key problems facing our age. There is no doubting the possibility of scientific breakthroughs that can profoundly affect all of humanity – and there are many examples. Particular universities can also play important roles in a crisis – as was shown during the Covid-19 pandemic in relation to vaccine development (University of Oxford) and monitoring and public information (Johns Hopkins University). These roles are real and valuable, and should not be belittled. Yet they are only one part of a complex puzzle. As is the case in society more broadly, the tasks of adaptation and regeneration in particular require localised work with human agency, and solutions simply cannot be given or imposed from the outside. The inescapable conclusion is that we need vibrant, quality higher education in all parts of the world.

## Time-bound

The final characteristic of climate change of relevance to our discussions is its relationship with time – most obviously, its urgency. Most commentators assert that radical action needs to be taken by governments and societies now, or we will reach the 'tipping point', at which climate change and its destructive impacts become rapid and irreversible, on account of the multiple feedback loops. The IPCC – which tends to have a conservative and less alarmist position on these matters – stated in its special report (IPCC 2018) that the world needs to convert entirely to renewables by 2050 to avoid a catastrophic temperature rise of 2°C. Given the cumulative nature of the impact of greenhouse gases, the later we leave our actions, the more difficult mitigation will be to achieve and the worse will be the impacts.

Political conflicts on the climate crisis have shifted onto this terrain of time. Whereas previously there may have been a debate as to whether anthropogenic global warming through greenhouse gas emissions was real, most stakeholders now accept this fact but differ in the urgency with which they approach it, and the level of sacrifice they are willing to make. In this way, those in the fossil fuel lobby attempting to resist change and maintain their profits have moved to a strategy of *delay* rather than *denial* (Thunberg 2024). Politicians – variously swayed by lobbying corporate interests or by the perceived self-interest of the electorate – often pay lip service to environmental measures while continuing environmentally destructive business-as-usual, pushing the responsibility along the timeline for future governments. Foot dragging and delay can also slip back into reactionary opposition. In this way, the Conservative government in the UK, which had expressed at least some commitment to addressing the climate crisis through policies on renewable energy, phasing out petrol and diesel cars, and improving household energy efficiency, was during the writing of this book, sending out populist messages of support for 'motorists', opposing '15 minute cities' and other environmental measures framed as being 'nanny state' and restricting individual freedoms.

One of the very complex aspects of time in relation to the climate crisis is that the impacts of our actions are felt only years, decades or centuries later. We are experiencing now the dark side of the accumulation of wealth made possible in the Global North by the Industrial Revolution; future generations will suffer from our excessive consumption today. This predicament presents a significant challenge to our customary moral sentiments: while we may adhere to a rational moral framework of obligation to human beings in the future point in time, we struggle to live that obligation as a felt experience. It is hard to feel strong moral ties to people in distant lands whom one has never met; it is even harder when those people have not yet been born.

These moral complexities present challenges to the university as they do any institution, community or individual. Just as individuals have to weigh up the competing interests of their immediate needs and wants with their commitments to ensuring a liveable planet for the whole of the global community of future generations, so universities can find themselves caught between the pressures of striving for financial security and academic excellence, and their public good contributions. Yet, as will be explored later in this book, the urgency of climate change also presents some distinctive challenges for the university. Scholarship and learning function in their own organic time and can rarely be rushed. The fruits of these processes frequently emerge in the medium or long term. As explored extensively in relation to recent emphasis on *impact* (social and economic, rather than academic impact), universities are not always well suited to bringing and showing immediate tangible changes in the outside world, and orienting themselves around that aim may in fact undermine their sources of value (McCowan 2019). A number of commentators have critiqued the pressures of time on contemporary academic work in the context of capitalism (for example, Shahjahan 2015; Facer 2023), with growing support for the 'slow scholarship' movement (Berg and Seeber 2016; Leibowitz and Bozalek 2018). These considerations do not absolve the institution from any responsibility, as many things can be done in the here and now. The important thing is to identify, acknowledge and plan on the basis of the diverse timings of different parts of university activity.

## Implications for the university

The characteristics outlined above present two major implications for the university. The first is, simply, that it is ideally placed – possibly even essential – to addressing climate change. As stated above, climate scientists are largely (though not exclusively) located in HEIs, and the lion's share of our knowledge on the issue stems from the work of universities. Universities and their staff also have a unique role in applying theoretical knowledge to practical questions of mitigation and adaptation, working together with government, private sector and civil society organisations. The changes in understanding and behaviour that are required in the population as a whole also mean the university is a crucial site of education, in conjunction with schools (Facer et al. 2020). Finally, as an institution that is oriented not only towards generation of knowledge, but also to questioning, debating and determining the basis of our knowledge, it is also well placed to address the epistemic challenges of the issue – and can intervene directly in that sphere through its teaching function.

Each of the five characteristics of climate change discussed draws the university into the picture. As the primary store and generator for humanity's knowledge, it needs to be engaged in addressing a crisis rooted in human society and human beings. As a space dedicated to deep and sustained thought, reflection and invention, from multiple perspectives, it is ideally suited to addressing the complexity of climate change. Contestation, although challenging, can be grappled with in the devolved and multifarious spaces for discussion found in universities, protected by the principles of academic freedom. The global role of universities is played out simultaneously through their international composition and reach, through the establishment of new institutions across increasingly diverse regions, and through online education and student mobility across national borders. The critical urgency of the climate crisis means that all societal institutions must contribute now, particularly when they are as potentially powerful as the university.

Nevertheless, all of these five characteristics, while making the university necessary, also raise significant tensions in the institution. The second implication then, relates to the ways in which universities should address the issue of climate change. The central place at the table given to the university in these debates does not necessarily mean that its traditional ways of working will be up to the task. Climate science involves interdisciplinary working that presents challenges to subject-based structures, traditions and taught courses (Leal Filho et al. 2018). Researchers are also forced to engage with political issues and currents in ways that may make them uncomfortable. Furthermore, and as argued in the chapters that follow, there will need to be a more holistic understanding of the workings of the institution, and of the interlinkages between research, teaching, community engagement and other functions.

The university is, therefore, essential to solving the conundrum of the climate crisis. It cannot do this on its own and must work closely and collaboratively with other spheres of society, but it is a key part of the solution. Nevertheless, the importance of higher education does not justify a 'business-as-usual' approach (Facer 2020). Profound transformations in the institution are needed to ensure that it maximises its potential and avoids the negative impacts on climate and human societies that it has had at times in the past. The rest of this book will focus on understanding what kind of transformation in higher education is needed and how it might take place.

## Notes

1 There is academic debate over the start date of the Anthropocene, with some placing it instead in the mid-twentieth century.

2 Following Bell (2021), this term is preferred to 'natural gas' as the latter was a phrase born of a corporate marketing strategy to promote its usage.

3 In the end the prize was never awarded.

# 3 Knowledge and climate change

The title of Amitav Ghosh's (2016) book on climate change – *The Great Derangement* – points to the incomprehensibility of humanity's current actions to future generations. At the centre of our 'derangement' is that we know about anthropogenic climate change – that it will bring suffering and ultimately extinction – and yet fail to do what is needed to stop it. This apparent lunacy has its roots in a number of factors. In part it can be explained by the actions (or deliberate lack of action) of corporations, states and social groups with vested interests in fossil fuels. Yet the problem lies also in the ways in which all human beings relate to the crisis. There are questions regarding the level of information and awareness that populations have, and disputes over the factual basis of climate change. There are also psychological considerations, such as the difficulties of dealing with risks that appear distant in time and space, and of sacrificing current comforts and changing hard-wired habits. Questions of climate change – its scientific basis and responses to it – have become embroiled in broader political currents, leading to resistance and opposition to climate action from certain segments of society. The notion of 'we' here is also problematic, with humanity divided in terms of its beliefs and values, unequal in power and wealth, and constrained in its ability to act in concert.

This chapter explores these epistemic dimensions of the climate crisis, and the multiple ways in which questions of knowledge are implicated in humanity's response. Naturally, as the knowledge institution par excellence, these questions are of central relevance to discussions of the role of the university. The chapter will have two main foci: the first on the quest for understanding the ways in which the climate is changing (measuring temperatures and levels of gases in the atmosphere, assessing environmental impacts and so on), and the second working out how to respond. Epistemically, these are quite different challenges. The first is to a large extent one of conventional science. It involves empirical observation, the building of hypotheses, testing and refining them – although its multidisciplinary nature challenges some aspects of the way science has conventionally been conducted. The second takes us into new territory. It does involve some basic science, along with applied science and technological development (for example, carbon capture, engine efficiency), but also involves social sciences, psychology and humanities, aiming to determine how to change the ways in which our societies are structured and how individuals behave. In this task, creating the conditions for an 'ecology of knowledges' (Santos 2015) is important, as we need different strategies to adapt and regenerate.

The above has focused on our collective knowledge, on the store of shared ideas and technologies that will help our species survive and thrive. But there are also questions of what we need to know individually. Clearly, not everybody needs to know everything about climate change – nor would it be possible, given the large range of relevant areas, and level of detail and expertise needed in each of them. Some degree of specialisation is justified also in the roles that we might take in combating the challenge. Yet at the same time, all people must have some knowledge and understanding of climate change, given the premise of this book that its causes are deep in the fabric of our societies and beings, and consequently that the transformation must also take place at all levels.

This chapter will therefore engage with the notion of knowledge in its broadest sense, including the formal knowledge of humanity that can be stored and communicated through language, as well as the embodied knowledge of people in their lived experience. As will be discussed further below, this task entails engaging with questions of belief, attitudes, values, skills and competencies, in addition to knowledge in the narrow sense (distinctions that are blurred and that manifest themselves in different ways in different cultures).

## The epistemic backdrop

As discussed in Chapter 2, concerns about the warming of the planet as a result of emissions of carbon dioxide and other greenhouse gases increased in the second half of the twentieth century, but only received formal international recognition with the creation of the UNFCCC in 1992. Since then, the world has taken only small steps towards dealing with the problem, through a combination of fragile international treaties developed through the COPs, national moves towards renewable energies, the development of less carbon-intensive technologies, pressure from popular movements and changes in consumer choices.

Given that climate change could make human life on Earth impossible in a relatively short timeframe, it is astounding how little action has been taken. Reasons for this inaction include: the deep embedding of carbon within the global economic system; the political influence of fossil fuel companies and the importance of fossil fuels to state power; the relative invisibility of climate change as a threat, particularly for those in cosseted higher-income communities; attachment to the luxuries of a carbon-heavy lifestyle, again particularly in higher-income communities; the difficulties of extending our moral imagination to future generations and to those geographically distant; the inertia caused by the overwhelming nature of the threat; and the difficulties of international cooperation on what is inevitably a global issue, as a result of economic and political competition between nations. As Ghosh (2016: 80) states:

> At exactly the time when it has become clear that global warming is in every sense a collective predicament, humanity finds itself in the thrall of a dominant culture in which the idea of the collective has been exiled from politics, economics, and literature alike.

While there are those who disbelieve, whether entirely or partially, in either the existence of anthropogenic global warming or its potentially catastrophic impacts, most do in fact accept their reality. According to a recent UNDP and University of Oxford (2021) study, 64 per cent of people worldwide believe that there is a climate emergency (a stronger belief than in the existence of climate change), and in most countries only 1 or 2 per cent of the population are opposed to all climate-related policies. The primary challenge, then, is how to convert this knowledge and understanding of the issue into concrete changes in individual behaviours, collective organisation and institutional structures of societies – and education is strongly implicated in this task.

The industrial capitalism that is the immediate cause of climate change was predicated on ontological and epistemological shifts that occurred in medieval and early modern Europe. In the first place, there was the dualism that – in contrast to traditional animism – separated humans from nature, imbuing only the former with soul, agency and true value, and legitimising the exploitation and even the extermination of the latter (Hickel 2021). In conjunction, there was the emergence and ultimate hegemony of a new conception of 'science', as a project to accumulate knowledge about the physical world that could be formulated into laws and used to predict and control the natural world. This ontological-epistemological complex underpinned both capitalism and colonialism, and made possible a level of intervention in and damage to the planet that far exceeded what humans in previous eras had managed.

Despite the dominance of this worldview, epistemic contestations continue. Traditional theocratic and aristocratic notions of truth and justice, Enlightenment notions of rationality and progress, and postmodern unsettling of objectivity and universality are historically layered but continue to coexist in contemporary societies – with varying configurations. The campaigns for the US presidential election in 2016 and Brexit in the UK counterposed the 'people' against the political elite – in these cases framed as the liberal, cosmopolitan and educated. Intertwined with this constructed tension was a distrust of specialists and academics, most famously encapsulated in the UK in Michael Gove's statement that 'people in this country have had enough of experts' (Mance 2016). These moves have led to a legitimisation of the questioning of official knowledge, science and experts, and an encouragement of counter-theories and conspiracy theories. While there was some recovery of recognition for science and experts during the Covid crisis, the pandemic also fuelled the spread of disinformation and conspiracies, along fault lines similar to those of climate change.

The rise of social media has intensified these trends, since they allow for rapid dissemination of 'alternative' truths, and the reinforcement of echo chambers, with users often confined to communities that reflect their own political views. Furthermore, social media platforms have increasingly been used as tools for the deliberate manipulation of the populace for commercial or electoral gain, making even more fragile the conduits to reliable knowledge.

Climate change is contested epistemically, but also axiologically, and touches deeply on questions of the good life. For some, the idea of a low-carbon existence is the ultimate sacrifice, the obliteration of all of the precious gains made by civilisation and technology over recent centuries. For others, it is a welcome opportunity for a less opulent, but ultimately more meaningful, healthier and more enjoyable life, freed from the noise and pollution of industrial society and the mental enslavement and manipulations of consumerist society. It is inevitable that those in these two camps will advocate for different strategies for addressing the problem: the former, technological fixes that will allow high-consumption lifestyles to continue; and the latter, emissions reductions that will require less wasteful forms of living.

These divergences are well illustrated by Ghosh (2016), who points to the dramatic contrast between the two important climate documents published in 2015: *Laudato Si'* and the Paris Agreement. While both endorse climate science, the former – written by Pope Francis (2015) – advocates, from a perspective of religious faith, the need for a radical transformation of human organisation and behaviour. The latter – a secular intergovernmental agreement – avoids criticising the paradigm of endless growth, and instead settles for a gradual voluntary change within the assumptions of our current global political economy.

Climate change also fosters political division over forms of societal organisation. Much of the resistance to and denial of the views of climate scientists, campaigners and advocacy organisations is because they are associated with collectivist organisation and restrictions of individual freedom. In this way, many climate change deniers brand environmentalism as communism through the back door, and even a smuggling in of a global super-government (Booker 2009; Morano 2018). While the transition to a low-carbon society does not necessarily entail a centrally planned economy, it is inevitable that such a significant societal shift will require some collective management and international cooperation.

Another aspect of significance is the emotional dimension of climate change. The strong embedding of values into the different positions on climate change already make it an emotive and potentially conflictual issue. Yet there is another significant element in the existential threat posed by the phenomenon. A number of commentators (for example, Lehtonen et al. 2019; Ojala 2016) point to the fear and anxiety caused by climate change, particularly in young people. While fear can be galvanising in some instances, in others it can be debilitating, or lead to a 'what the hell' attitude and a doubling down on destructive actions. These affective and psychological dimensions must be acknowledged and engaged with in responses to the climate emergency.

Before moving on to consideration of the question of how individuals acquire knowledge and understanding of climate change, and convert it into action, there will first be a discussion of the collective development of knowledge through research and scholarship.

## Understanding climate

### Climate science – is there a consensus?

One of the central tensions in our knowledge of the climate revolves around consensus. Scepticism in science is a virtue and it is not an overstatement to say that, without scepticism, iconoclasm and rebellion, there would be no modern science. And yet, on the issue of climate change, the academic community has been at pains to show its unity and prove the existence of a consensus. This position has to a large extent been forced on scientists by the orchestrated attempts from the fossil fuel lobby to undermine certainty in the public eye. Yet it remains an uncomfortable position for those committed to the building of knowledge through open enquiry, embracing of doubt and the presentation of unfiltered, complex reality to the general public – on the basis of the Popperian principle of falsifiability. Perhaps more than any other issue, climate change has brought to the surface the tensions between scientists' commitment to truth and their responsibility to humanity.

In response to the 'merchants of doubt' (Oreskes and Conway 2010), a number of studies have been carried out on the level of consensus among climate scientists. Most of these analyse the abstracts of published papers to assess whether they express implicit or explicit agreement or disagreement with anthropogenic climate change. Naomi Oreskes's (2004) early study showed that, while not all of the 928 studies analysed explicitly endorsed the consensus position, none explicitly rejected it. A later study by Cook et al. (2013), analysing 11,944 papers, put the consensus level at more than 97 per cent of those papers expressing an explicit view, and figures of more than 90 per cent consensus have been confirmed elsewhere (Cook et al. 2016; Lynas et al. 2021). Powell (2019) even asserted a consensus of 100 per cent on publications analysed from 2019. Other studies (for example, Myers et al. 2021) have involved surveys of active scientists and have supported the analysis of publications in asserting over 90 per cent endorsement of the consensus.

The work of providing a unified scientific voice on the issue is fulfilled most prominently by the Intergovernmental Panel on Climate Change (IPCC). This body was established in 1988, bringing together a large group of experts to review existing climate research and provide periodic assessments reports (in cycles of between 5 and 8 years). These assessment reports cover a range of different areas, with the three principal working groups covering climate science, impacts on human and natural systems, and mitigation actions. The IPCC does not carry out its own research but functions as a clearinghouse for the huge volume of climate science produced globally, providing a measured and cautious voice distinct from the more explicitly politicised demands of activists.

Yet even with the IPCC's work, tensions over science and public perception have not been completely resolved. As argued by Almassi (2012), trust in scientists is essential in the case of climate change, since it is unfeasible for most members of the public to have direct knowledge and understanding of the science, and in most cases we do not have conclusive proof of anthropogenic climate change through our everyday experience. However, that trust has been shaken, not only by the general epistemic destabilisation discussed in the previous section, but also by the concerted efforts of the fossil fuel lobby to undermine confidence in the science. Those with vested interests in fossil fuels exploit the existence of unsettled questions to cast doubt on the need for action, while efforts by the scientific community at presenting a unified voice are branded conspiracies aimed at silencing dissenters. The 'tuning' necessary in computer models of climate is an easy target for accusations of manipulation (Koonin 2021). The so-called 'Climategate' incident – involving the hacking of emails and data from the University of East Anglia – fuelled these allegations: although in fact taken out of context, the academics' comments supposedly indicated the manipulation of research findings.

The tensions between, on the one hand, academic integrity in terms of robust and transparent methods and faithful representation of findings, and, on the other, responsibilities to society and humanity and the associated moral and political commitments, are common to all areas of academia, yet are particularly fraught in climate change where the stakes are so high. Given the huge difficulty in changing individual behaviours and government policy, a feeling of uncertainty, however small, may be enough to undermine meaningful change. The following section will explore further the concerted attempts to destabilise consensus.

## Climate denial, distraction and delay

During the writing of this book in 2023, my Twitter feed started to fill with #climatescam posts reacting to various environmental issues such as the '15 minute city', climate protests, or the wildfires that gripped Greece and other countries during the summer, and even weather reporting (the supposed manipulation of colouring on television images to intensify feelings of heat panic). Many of these tweets questioned the very existence of climate change, arguing, for example, that carbon dioxide is good for plants, so why would we not want more of it? While not always putting its head above the surface, opposition to climate action still lurks underneath.

According to climatologist Michael E. Mann (Mann and Toles 2016), there are 'six stages of denial', corresponding to the following positions:

1. 'It's not happening!'
2. 'Ok, it's happening . . . But it's *natural*!'
3. 'The problem is self-correcting anyway.'
4. 'And it will be *good* for us!'
5. 'It's too late or too expensive to act.'
6. 'We'll find some simple technofix anyway.'

Not all sceptical positions, therefore, constitute absolute denial of anthropogenic climate change. Coady and Corry (2013) similarly argue that there are four areas of potential contestation: (i) the world is getting warmer; (ii) it is caused by humans; (iii) this is bad; and (iv) we should do something about it. Only the first two (corresponding to Mann's points one and two above) actually constitute direct denial of the existence of climate change (and the scientific consensus outlined in the previous section). The others relate to its impact, its desirability and the responses that we should make to it. All of these in fact need to be brought into a discussion of climate denial, more broadly conceived, as they all serve to undermine humanity's ability to address the issue and are often merged and conflated in deniers' statements.

The term 'denial' is clearly a loaded and emotive one – given its associations with Holocaust denial, and its Freudian overtones – and unsurprisingly it is not used by those doing the denying (they usually prefer the term 'sceptic'). Coady and Corry (2013) distinguish between *scepticism* (not believing one or more of the four claims outlined above, possibly because of insufficient evidence) and *denial* (actively believing one or more of these claims to be false); yet most often the usage of these terms is rhetorical, intended to exude either positive or negative qualities.

As shown by the UNDP and University of Oxford (2021) survey mentioned earlier in this chapter, the vast majority of people worldwide accept the existence of anthropogenic climate change and recognise its seriousness, yet there exists a minority that do not – and it is often very vocal. There are a variety of reasons why individuals and groups resist the facts of anthropogenic climate change, including attachment to a high-carbon lifestyle, political opposition to environmental groups, and affiliation with far-right groups and parties. These propensities, however, have been exploited and intensified by the deliberate efforts of corporations.

The book and documentary *Merchants of Doubt* by Naomi Oreskes and Eric Conway (2010) shows the history of corporate obfuscation of scientific evidence that threatens profit-making activities. Despite knowing the health risks of smoking from as early as the 1950s, the tobacco industry adopted various techniques of distraction in order to avoid regulation; these same techniques have been adopted by the fossil fuel companies to protect their interests in the context of the developing science on climate change. Alice Bell's (2021) history of climate change in fact shows the deeper roots of corporate efforts to support a carbon economy: the early automobile industry coined the term 'jaywalking' in order to naturalise the predominance of cars on public highways and criminalise pedestrian access. Climate denial campaigns have mobilised a small number of recognised scientists who are climate sceptics (such as Fred Singer and William Happer), along with specially created think tanks and alliances with neutral-sounding names, such as the CO2 Coalition (formerly the George C. Marshall Institute). The political lobbying is supported by efforts to change public opinion through websites (for example, Climate Depot, Watts Up With That?), books (for example, Inhofe 2012; Morano 2018; Booker 2009) and social media activity.

The different types or strengths of denial outlined by Mann and Toles (2016) and Coady and Corry (2013) show themselves clearly in these activities. What we might call 'full-fat' denial involves questioning or dismissal of anthropogenic climate change outright: examples here include Marc Morano (see *The Politically Incorrect Guide to Climate Change*, 2018) and the late Stephen Booker (see *The Real Global Warming Disaster*, 2009). This position is becoming less common, and efforts have moved on to the next two types. 'Semi-skimmed' denial – such as Bjorn Lomborg (see *Cool It*, 2007) – accepts that temperatures are rising on account of the burning of fossil fuels, but does not accept that the impacts will be as negative as portrayed, and argue that there may even be some positive benefits. 'Skimmed' denial accepts the existence of climate change and its negative consequences, but questions the settled nature of the science, and therefore of humanity's response. Steven Koonin's book *Unsettled* (2021), for example, contrasts with the works mentioned above as it takes a more measured approach. Written by a prominent scientist, it does not deny climate change, but argues that climate scientists have not been fully honest with the level of doubt around the topic. In their eagerness to change public opinion and galvanise action they have papered over the cracks, and thereby undermined their own scientific credentials and ultimately public trust. While it is hard to deny the need for scientific honesty, Boslough (2021) counters that what Koonin and others have done is to create a strawman, an easy target that does not in fact correspond to the reality: climate scientists have never in fact claimed that all aspects of the topic are settled or that there are not any uncertainties in their calculations and predictions.

But rather than fitting neatly into one of these three boxes, or clustering around different stages on Mann and Toles's (2016) scheme, proponents of climate change denial often combine positions, leading to confusions and contradictions in their arguments. Booker (2009), for example, moves between critiques of the methodology of the 'hockey stick' study (Mann et al. 1998) and the inefficiencies of renewable energy (with a particular hostility towards wind turbines), along with attacks on the United Nations, European Union and big government. One thing that remains unclear from the climate denial contingent is what the reasons might be for scientists, politicians and activists in concocting this great conspiracy of climate change. The three reasons given most commonly are: (i) deliberate smuggling in of other political agendas, in particular socialist, egalitarian and state-centric policies; (ii) self-interest – for example, scientists wanting to obtain research grants and awards; and (iii) mass hysteria, in the same vein as the documented instances in medieval and early modern Europe. But none of these motivations seems sufficient explanations for initiating such a giant endeavour, or correspond to the way the global climate movement has operated in practice.

Yet, despite these inconsistencies in argument and the general lack of evidence to support their positions, climate change deniers have been effective in sowing the seeds of doubt and in holding their own in public fora. As explored by Marshall (2014), spokespeople for the climate denial lobby, such as Marc Morano, have proved themselves to be more adept at the public communication game than climate scientists, and have had some success in undermining the scientific consensus. While it may be a minority of people who fully deny climate change, even a small amount of doubt introduced can be harmful, given the difficulties of achieving a radical shift in our individual and collective lifestyles and worldviews. Now the fossil fuel lobby has largely moved from denial to *delay* and *distraction*, a small amount of doubt can be highly conducive to buying time to carry on making profits during a slow transition.

To a large extent, denial of climate change and rejection of the policies designed to address it have little to do with the facts of the matter but are more an emblem of allegiance. Marshall (2014) argues that, far from being swayed by the empirical evidence, the right-wing groups that deny climate change hold rejection of the science as a badge of honour, forging stronger affiliation with the in-group. Climate change denial is, therefore, strongly linked in certain contexts with a range of other causes, such as individual freedoms, the free market, small government and the rejection of metropolitan liberal elites. Yet denial is not limited to these right-wing groups. As Norgaard (2011) shows, even communities which have knowledge and experience of climate change engage in an unwitting 'socially organized denial', by failing to make connections to human actions – as will be explored later in this chapter.

There are significant implications of climate change denial for universities, both in relation to science and its public communication, and to the curriculum and processes of teaching and learning. These debates will be addressed in the chapters that follow.

## Knowledge in climate responses

It might be assumed that, once the scientific knowledge on the causes of climate change and its impacts exists, then the epistemic problem would be solved, that all we would need to do is 'act'. Unfortunately, it is not so simple. First, the complexity of the climate system and the human systems interacting with it are such that the appropriate action does not necessarily emerge from a good understanding of the causes and impacts. For example, while we might identify accumulation of carbon dioxide through burning fossil fuels as the primary target, stopping burning fossil fuels requires a whole new body of knowledge around energy usage in society, alternative energy sources and strategies for making the transition. New technologies are required for bringing about other kinds of solutions to the greenhouse gas build-up, such as carbon capture.

Moreover, mitigation is not the only challenge. Changes already under way require us to adapt, and the long-term establishment of a sustainable planet also requires us to regenerate. Adaptation and regeneration are mostly local questions, contextualised and specific to the unique geographical, cultural, political and economic conditions of each place, and so require forms of knowledge that are grounded in those places.

Finally, there is the 'we' in 'what knowledge do we need'. The globe is characterised by various forms of inequality and disparity, and knowledge is not equally at the service of all. Furthermore, there is the question of individual acquisition and uptake of knowledge, and the relationship with behaviour: it is not just a requirement of justice that knowledge should be available for all of humanity, the very solving of the climate conundrum depends on it. The educational challenge, therefore, is how to ensure that all people understand the climate crisis and are also equipped to do something about it.

This section will, therefore, address the question of the knowledge that we need to respond to the crisis, and how it might be generated, communicated and distributed. The first part will focus on knowledge production (research, scholarship, application of knowledge and innovation), and the second on the communication, distribution and acquisition of knowledge by the broader population.

## Generating knowledge

Knowledge production is highly incentivised in many universities. Academics are normally committed to their disciplinary areas and have intrinsic motivation to enquire and communicate about their chosen areas. Academic career incentives reward success in grant funding and publication. International university rankings are also heavily weighted towards research, particularly publication in elite journals and volume of citations. Most higher education systems also have academic freedom as a core value and allow researchers space to pursue their own lines of enquiry, even when these counter commonly accepted views. These factors have ensured that, in the relatively short space of time since the establishment of the IPCC in 1988 (although building on the scientific groundwork of earlier decades and centuries), the academic community has developed extensive knowledge on climate change, with an estimated 10,000 papers a year on the topic (Koonin 2021).

There have also been incentives to develop applied knowledge and technological innovation. Following the model of the entrepreneurial university (Marginson and Considine 2000; Shattock 2009), universities have promoted partnerships with industry, established spin-off firms and registered patents in an effort to boost their third-stream income, while simultaneously generating new knowledge and products. These efforts to find climate solutions have been encouraged by prizes and promises of investment from prominent entrepreneurs such as Richard Branson and Bill Gates.

Nevertheless, there are constraints on what knowledge is produced and how it is distributed. The conducive conditions outlined above are in reality restricted to the well-funded and prestigious institutions of a few countries in the Global North: of the estimated 90,000 HEIs in the world (MacGregor 2022), only some 2,000 are even listed in the international university rankings, and only some of those have the funds and infrastructure to carry out complex scientific projects. For those institutions that are less well funded and with lower public recognition, even if they can carry out high-quality research, their opportunities for publication and dissemination are constrained, particularly if they use a language other than English.

There are also questions of public and private. The influence of higher education on society is strongly dependent on the public good nature of knowledge (Marginson 2011, 2018; Singh 2012; Unterhalter et al. 2017). While it is possible to temporarily retain knowledge as a private good (for example, through restrictions on intellectual property, or through a selective intake of students), over time, this knowledge tends to seep out and become generally available. This is not to say that the knowledge produced is of equal benefit to all, or that all people are equally well positioned to take advantage of it: economic and educational inequalities ensure that there are significant disparities in the use made of knowledge, even when it is publicly available. Furthermore, the cost of access to new technologies in a market system can restrict its use to the few, both internationally and within countries.

As discussed in an earlier treatment of higher education and the SDGs (McCowan 2019), international rankings reward research excellence, but give little recognition to the impact of that research on society. The developmental model of university – which is oriented primarily towards positive impacts on society, particularly in relation to the most disadvantaged segments – is still alive but struggles for space in the context of status competition and marketisation. Much research on climate may not find an outlet in the top journals but may nevertheless bring significant advances in understanding and positive impact on the ground.

There are also questions of disciplinary divide. Most commentators on the topic assert the need for disciplines to work together in order to address the complexities of climate change. This combination might be multidisciplinary (different academic areas working together but remaining separate), interdisciplinary (different academic areas entering into dialogue with each other, and modifying each other through their joint working) and transdisciplinary (going beyond disciplines altogether). Yet combining disciplines is a great challenge, on account of the distinct sets of truth criteria, methodologies and values underpinning each, and despite some efforts on the part of institutions and funders, institutional structures and incentives remain disciplinary.

It is also important to consider how these disciplinary perspectives are balanced. A study by Overland and Sovacool (2020) showed that only 10 per cent of funding for research on climate change mitigation goes to the social sciences and humanities, with the vast majority going to science, technology, engineering and mathematics (STEM)[1] areas. Given the roots of the crisis in political, economic and cultural spheres of society, this disparity needs to be addressed. The growing field of environmental humanities, and initiatives such as the UNESCO BRIDGES programme, are making steps to address this imbalance.

Most obviously in relation to adaptation and regeneration, but also for mitigation, research must involve non-university communities (Rapley et al. 2014) – as will be explored in greater detail in Chapter 9. The notion of 'co-production' – the joint creation of knowledge between researchers and other citizens – has become something of a buzzword in recent years, but in practice is rare. Co-production involves two interconnected elements: in the first place, the involvement of different actors (for example, university-based researchers and members of the local community), but also the joining together of different modes or traditions of knowledge and epistemologies.

Boundaries between different knowledge traditions – for example, Western/non-Western and mainstream/Indigenous – are not watertight. In so-called Western cultures there are strong historical influences of non-Western knowledge, as well as diverse knowledge traditions (for example, intuition and tacit knowledge – Binagwaho et al. 2022). There remain very few Indigenous communities without contact with other groups, and as a result, the circulation of hegemonic knowledge traditions is almost universal. Nevertheless, there are differences still evident, in relation both to what is known and how it is known, and many are bound up also with the languages in which they are expressed.

Why should these differences be of significance to climate action? Mainstream Western science has proved itself to be remarkably adept at identifying the principles of the physical world around us and using that understanding as the basis for technological development. Yet given the extraordinary changes in humanity's perceptions of reality over past centuries and millennia, only extreme hubris could maintain that the knowledge currently held in Western academia is the final or the only version of things. Epistemic humility is needed to understand that there are other valid forms of knowledge and ways of understanding and living in the world.

Knowledge production and co-production, therefore, need to engage with what might variously be known as local, traditional or Indigenous forms of knowledge. There are three bases for this principle. In the first place, it could be seen as a question of justice, a right – in granting to all communities the respect due to their languages and knowledge traditions. In this way, there has been attention of late to questions of epistemic injustice: for example, the work of Fricker (2007), distinguishing between 'testimonial' and 'hermeneutical' forms – the former referring to the supposed unreliability of certain groups as sources of knowledge, the latter to the lack of the conceptual tools to understand their distinctive forms of knowledge. Second, it can be seen that there is intrinsic value in different knowledge traditions – that is, they have value in themselves, in that they provide either valid or desirable perspectives on reality. Finally, there may be instrumental justifications. The last of these is the most commonly cited – for example, in the use of Indigenous knowledge of plants for medicines, or of weather patterns for disaster risk reduction. An overemphasis on this last point can lead to problems of *instrumental* co-production – how Indigenous knowledge can be used in ways that select only those parts that fit with Western science (Goldman et al. 2018).

There are strong arguments and many advocates for the inclusion of Indigenous, local and traditional knowledge in climate adaptation generally speaking (Byskov and Hyams 2022). In relation to higher education specifically, engagement with these diverse forms of knowledge is a challenge, on account of deeply entrenched epistemic norms. It is important to emphasise that accommodating this diversity does not mean displacing mainstream academic knowledge but placing it in dialogue with other forms of knowledge (Santos 2015; 2017).

Building an ecology of knowledges in the university is no mean feat. Even maintaining the use of multiple languages has proved too much for higher education systems, which have in most cases remained monolingual in the context of plurilingual societies, often retaining colonial languages that are not widely spoken and increasingly moving towards publishing in English. The greater demands of genuine epistemic pluralism – beyond a more superficial valuing of Indigenous knowledge of plants and weather, for example – require a major shift in our HEIs, one that may call for the forging of a whole new institutional type (McCowan 2021b). The steps that institutions must follow in the interests of fostering epistemic pluralism will be outlined in the chapters that follow.

## Acquiring and using knowledge

Different kinds of global crisis require different forms of knowledge distribution. The threat of an approaching meteor can be staved off by scientific understanding and technologies developed by a small group of people in a single location, with their successful deployment – diverting or destroying the meteor – benefiting all people in the world. While some certainly argue that a solution of this kind can be found for climate change, it is highly unlikely, given the need for changes in consumption patterns, lifestyles, relationships to the natural world, distribution of resources and so forth. While some individuals and communities may have specialised knowledge and roles, all people need some knowledge and understanding of the phenomenon.

The first question facing us is: what exactly do people need to know? It is hardly feasible that everyone has as detailed and sophisticated a grasp of all the issues as a professional climate scientist. Yet everybody needs enough of an understanding of greenhouse gases, the functioning of the atmosphere, the impacts of temperature rises and so forth, that they are able to make changes in their own lives, and to evaluate and either support or oppose the actions of corporations and the policies proposed by their governments. The knowledge needed will involve not only scientific evidence but also the political, economic and cultural factors that determine, facilitate and constrain climate action: not only propositional knowledge but also experiential learning, skills and values.

There is, therefore, learning needed by all members of society, but there are also individuals who have specific roles in relation to climate change and who, as a result, require particular forms of learning. These will include professionals working directly with environmental or social questions relating to climate change, and scientists and scholars researching the phenomenon, both of whom are most likely to be trained in universities. This dual role for the HEIs – in forming professionals and educating citizens – has particular implications for the way climate change is addressed in the curriculum, as will be explored in Chapter 7.

The second question is: how should that knowledge be communicated and acquired? Public communication has been at the centre of the struggle around climate in recent decades, pitting scientists and activists against lobbyists and think tanks aiming to disrupt their messages, and against governments that are slow to move and preoccupied with other issues. As argued by George Marshall (2014), the climate science community has on many occasions been outmanoeuvred in the public communication battle. Academics are having to shift from 'throwing their papers over the wall' (Rapley 2023) and trusting that the public will take up their ideas, to developing skills of accessible and persuasive communication. These questions of public communication of climate, which are dealt with amply elsewhere (for example, Marshall 2014; Rapley et al. 2014), will be taken up again in Chapter 9.

Nevertheless, even fostering knowledge within the university space encounters significant barriers. Knowledge and understanding exist at different levels of depth and have a complex relationship with action. The vast majority of people in the world now know about the existence of climate change and recognise its dangers (UNDP and University of Oxford 2021), and yet the global community is not doing what it needs to in order to address the issue. As Norgaard's (2011) ethnographic study of a small town in Norway shows, it is possible to be highly educated and well informed about climate change, and still live in self-denial of the need for change.

Educational aims have conventionally been split into knowledge, skills and values. While the term 'knowledge' can also refer to acquaintance or to know-how or procedural knowledge (closer to 'skills'), generally it is used to refer to the factual or propositional. To the knowledge element we can add *understanding*, which is the deep absorption of knowledge so it is grasped and becomes meaningful for the individual in question. In the case of climate change, this will include, among other elements, knowledge and understanding of the evidence surrounding changes in temperatures over time, the causes of anthropogenic climate change in greenhouse gas emissions, the aspects of human behaviour that lead to greenhouse gas emissions and ways of absorbing carbon from the atmosphere. The knowledge component is the focus of the widely used Sulitest,[2] a tool designed to assess sustainability literacy among students and the general public.

The term 'skills' refers to practical abilities to do things, rather than just to know things. A distinction is often made between hard skills – technical or profession-specific skills – and soft skills, which are generic ones related to personal and interpersonal abilities such as teamwork. In relation to climate change, skills needed may be those of research and analysis, but also political skills of advocacy, organisation and campaigning, and those related to environmental protection and use of new technologies. Currently in vogue is the idea of twenty-first-century skills (IT literacy, creativity, collaboration and so on), seen as essential to address the challenges of our complex post-industrial societies, including climate change. UNESCO (2006: 21) provides the following list of skills seen to be essential for addressing sustainable development:

> [C]reative and critical thinking, oral and written communication, collaboration and cooperation, conflict management, decision-making, problem-solving and planning, using appropriate ICTs, and practical citizenship.

The development of these transferable skills has not always been seen as the remit of the university. Those skills most naturally associated with higher education are those of intellectual enquiry, including textual analysis and inference, bibliographic searching, empirical data collection, conducting scientific experiments, as well as debating and public speaking. More specific skills of an academic nature are developed in each discipline. Yet the massification of higher education in the twentieth century has led to increasing numbers of vocational courses, with more practical job-related skills, and an emphasis on employability. These trends have opened up the range of skills in the purview of higher education, although not without resistance.

Finally, there are values. Values relate to different spheres (including aesthetic), but those most relevant here are the political and the moral. We may see values as having intrinsic importance, but they also have instrumental relevance in relation to climate change since they shape beliefs, attitudes and behaviours. Values relating to climate change include those relating to the natural world (what Orr (1994) calls 'biophilia'), but also those relating to social justice, equality and liberty. Since greenhouse gas emissions are the result of social, political and economic organisation, reducing them necessarily entails changing the organisation of society and the distribution and use of resources. Global disparities bring into play questions of climate justice and environmental justice more broadly, the need to develop a 'moral imagination' (Reimers 2021) and the need to understand the centrality of inequalities to the current crisis.

Knowledge, skills and values are all essential for human beings to engage in climate action. Literature on the topic has shown convincingly that simple awareness-raising and development of knowledge about climate change is insufficient for ensuring action and changes in behaviour (Anderson, A. 2012; Facer 2020; Facer et al. 2020; Monroe et al. 2019; Rousell and Cutter-Mackenzie-Knowles 2020; Stevenson et al. 2017). Research by Ojala (2016) also shows the necessity of engaging with climate change as an affective or emotional issue, not just a cognitive one. Others (for example, Nussey 2021) have argued for the importance of cultural, artistic and aesthetic engagement with issues around climate. The reality of university provision may be somewhat different, however: the UNESCO (2019) analysis of country submissions under the UNFCCC reported that, at the tertiary level, 75 per cent of courses focused on knowledge, 0 per cent on skills and 25 per cent on socio-emotional skills.

We might question the divisions between these three elements of knowledge, skills and values. The fragility of the distinctions between them can be shown by their usage in different languages, even among relatively similar languages such as those in the Indo-European family. In Spanish and other Romance languages, for example, there are two words for knowledge, which in some usages can distinguish between more theoretical and formal knowledge (*conocimientos*), as opposed to more practical or day-to-day knowledge (*saberes*). This distinction starts to blur the line between knowledge and skills. We could also argue that it is hard to extract values entirely from either knowledge or skills. The acquisition of knowledge can lead to the development of certain values: prejudice against migrants and foreigners may be based on misconceptions (for example that they are 'stealing our jobs' or 'sponging off the state') that can be shifted with exposure to the actual evidence; meeting and interacting with people from a hostile outgroup (knowledge as familiarity) can lead to seeing their common humanity and changing feelings towards them. Likewise, skills – particularly complex sets of skills involved in trades and professions – often come with an embedded set of values towards the practice itself, such as pride or commitment to the high standards and aesthetics of, say, glassblowing or landscape architecture. Similar debates are played out in relation to critical thinking, which is, at first sight, a skill or a set of skills but is hard to disentangle from the disposition to be critical (Schendel et al. 2023).

The objectives of learning are often framed in terms of 'competences' or 'competencies' (for example, Burandt and Barth 2010; Barth et al. 2007; Pérez Salgado et al. 2012). Competences are another example of the merging together of aspects of knowledge, skills and values, signifying the combination of attributes necessary for success in a particular activity or profession, or generally in life. This idea has become popular in the field of education for sustainable development – for example, in the notion of *Gestaltungskompetenz* (literally 'shaping' competence, or capacity for transformation) in the German language debates. In the field of climate change, there have been attempts to define competences in this way, with Burandt and Barth (2010), for example, putting forward the following set of four:

1. analysing multiple networked, complex problems of (non-) sustainable developments and the perspectives of sustainable changes;
2. dealing with uncertainties and thinking proactively;
3. using, shaping, handling and sharing different sets of information and knowledge;
4. assessments, ethical orientations and proactive thinking in order to secure a capacity to act.

Competencies are seen to be a progressive approach to framing the curriculum, since they move beyond inert knowledge content and rigid disciplinary approaches and towards multidimensional and real-life abilities, combining knowledge, skills and attitudes. Yet they bring with them some other problematic elements. One of these is that they represent an exogenous frame for learning: they determine what individuals should be able to do and how they should act on the basis of tasks or jobs that need completing. That may be entirely appropriate in assessing individuals for a particular form of employment but fall short of a holistic vision of education based on learner agency. Competence-based frameworks do not always contain a sufficient element of criticality and reflection, and in this regard can be more akin to *training* than *education.* (This risk is real in the field of climate change, with technical approaches bolstered by some of the international frameworks – Nussey 2021.) In addition, if they are viewed as a rubric to be completed, they can ignore the continuous dimensions throughout the whole of life: one does not reach a point at which one has sufficient imaginative thought or problem-solving and can stop developing them.

There are uses of competencies that avoid the above issues and put forward more open conceptions of interculturality, empathy and interdisciplinary work, on a lifelong basis. Nevertheless, even in these instances, it is not clear whether the notion addresses adequately the value dimension – for example, whether we should ever describe as a *competence* a person's commitment to protecting their community from a mining project or campaigning for racial justice, or even treating their work colleagues fairly and respectfully on a day-to-day basis.

Another distinction is useful here, that between learning *about*, *for* and *through*, one that has been used in relation to education for sustainable development (for example, Sauvé 1996), as well as citizenship education and human rights education (Kerr 1999; McCowan 2009, 2013). In the context of climate, this triad distinguishes between, in the first place, learning *about* climate change: the gaining of information on the topic and understanding of the debates surrounding it, and therefore linked strongly with the element of knowledge outlined above. The second – learning *for* climate change – is primarily related to skills, although may involve elements of knowledge and values as well, in developing the capacity to act: either directly in climate change mitigation or adaptation, or indirectly through campaigning, mobilising and influencing. The third, learning *through* climate change, involves experiential learning, taking place not in formal educational settings such as the classroom, but through activities relating to climate change in broader society, such as environmental projects, political action or community engagement work. This final element may also involve acquisition of knowledge, skills and values, but has some unique characteristics in that it goes beyond simulations to show the messiness and blockages – as well as the inspiration – of real-life settings. Nevertheless, these real-life settings are unpredictable and not always accessible, and they may not provide the best location for the development of knowledge and skills, so simulations within the university space will also be vital. There will always be a place for learning about and for, in addition to through.

There are strong arguments to suggest that what students should learn about climate change should, to some extent at least, be determined by what they want to learn about it. At the higher education level, student agency in relation to learning is both possible and highly important, and particularly in an area such as climate change in which they may already have extensive engagement. Existing evidence shows that university students are already strongly interested in and committed to resolving climate change – a recent survey in the UK, for example, has shown that 90 per cent of students are fairly or very concerned about it (SOS-UK 2021). Coercion in this area appears to be unnecessary, as well as undesirable. Monroe et al. (2019) conducted a systematic review of the effectiveness of climate change education: while most of the studies identified were at school level, 11 of the 49 were in higher education. The main points emerging as key to success in these cases were: '(1) The programs focused on making climate change information *personally relevant and meaningful* for learners. (2) The activities or educational interventions were designed to *engage learners*' (original emphasis, p. 9). These findings show how important it is to contextualise the content in ways that are relevant to students – either to their actual lives or their course content – along with the use of innovative and experiential pedagogy.

This brings us to one of the most complex questions of climate change education: the extent to which it should instil a particular set of knowledge, skills and values, or alternatively leave students to critique them, and allow them the autonomy to adopt alternative views. At first sight, there appear to be strong reasons for promoting a predefined set. Climate change is a crisis of potentially unparalleled proportions, threatening to wipe out humanity, so is hardly a 'take it or leave it' topic. Furthermore, there is a substantial body of coherent evidence on the trajectory, causes and impacts, leading to clear implications in terms of reduction of emissions of greenhouse gases. From the urgency, the gravity and the need for action, it would seem to be possible to derive a clear educational plan in terms of what all people must know and be able to do – and indeed this has been the basis of much educational action in relation to climate change.

However, while not dismissing the above out of hand, certain caveats are necessary with this position. First, there are still a number of unresolved empirical questions relating to climate change that require further investigation: while anthropogenic global warming may not be in doubt, our understanding of its speed, impacts and causes are constantly being refined, and learners should be aware of the ongoing processes of enquiry. Second, the response needed from the global community to this critical challenge is far from straightforward, and even among those who are both convinced of the reality of the climate crisis and committed to addressing it there are a range of reasonable positions, involving different combinations of energy efficiency, changes in consumer behaviour, regulation of corporations, geo-engineering, absolute reduction of resource use, redistribution and paradigm-shifting. It would be entirely inappropriate to present students with a single solution to this conundrum. Third, and most crucially, climate change education at the end of the day is still education, so it must be part of the general process of developing learners' own understanding, enquiry and critique, rather than unquestioning absorption of information or unreflexive training or conditioning (Jickling and Wals 2008). After all, it was only through this kind of critical scrutiny and challenging of existing beliefs that breakthroughs in climate science and political action were possible in the first place.

As stated by Facer et al. (2020: 3), SDG 13 on climate action:

> . . . implies that there is a 'settled' and somehow agreed framing of what constitutes climate change mitigation, adaptation and resilience – and implies that the function of education is to

> 'promote' this settled understanding, assuming a 'deficit [sic] knowledge approach that arguably seeks to fill in the so called climate knowledge voids in learners' heads, rather than to negotiate mitigation, adaptation and resilience measures as appropriate in each setting.

A predefined body of knowledge around climate change, or even a set of competencies, therefore, while playing a role, will never be sufficient as an educational programme in climate change. Ultimately, learners need to develop agency, to engage in enquiry, construct their own perspectives and generate new knowledge. Furthermore, the need for criticality – and also imagination and creativity – lead us towards a model in which, instead of predefined content delivery, students have ample opportunity for self-directed and peer learning. These distinctions and considerations will inform the reflections in the subsequent chapters on whether and how the university should incorporate climate change into its educational programme.

## Notes

1 Some forms of STEM research carry greater costs on account of the equipment and logistics involved, but this explains only part of the disparity.

2 https://www.sulitest.org/en/index.html

# 4
# Modalities and pathways of impact

References to higher education in international declarations on climate and sustainability (if it is mentioned at all) tend to assume that its function is to educate the youth – or more narrowly to train them for specific jobs. Doubtless, the educational function of these institutions is the most prominent, and perhaps important, one. Yet to assume that universities function like primary or secondary schools for young adults is to misunderstand them badly.

In the first place, HEIs have a diverse array of functions beyond teaching and learning. In most cases, they conduct some research, whether that is of a formal and public nature, through projects, labs and external funding, or more personal, through enquiry and scholarship. Often, they are engaged in the application of knowledge in more practical ways, through technological innovation, consultancy, secondments and evaluations for governments, companies or civil society organisations. They also serve as archives of knowledge, traditionally through libraries of books and journals and more recently through digital collections. Larger institutions may provide services for the public, such as hospitals and legal clinics, and in recent years it has been increasingly popular to have spin-off businesses. HEIs are often residential communities with the full array of services for their populations. The increasing complexity of institutions through the centuries led Clark Kerr (1963) to coin the term 'multiversity' to describe this array of functions.

This multiplicity of activities means that the university is not only a space for conveying ideas about climate change (for teaching it), but also for creating knowledge about it (collecting evidence, interpreting, theorising), for storing that knowledge and making it available to the public, and for applying it practically to society. This chapter aims to map these diverse functions in order to build a better understanding of the institution's contributions to climate action. It charts the different kinds of influence that these functions have on climate via a series of pathways – passing through the people who come into contact with universities, broader societal trends and the natural environment – and the ways in which these pathways interact. Some examples are given here, although fuller exploration of these different spheres of action is provided in the chapters that follow.

Like all frameworks, the one provided here is a painted portrait, highlighting certain characteristics of the subject and obscuring others. And, like all analysis, it artificially divides and separates what might otherwise be understood as an indivisible whole. But in order to explore and understand phenomena some kind of map is needed, even if we discard it at a later point. As will be drawn out more fully in the conclusion, this kind of scheme can help us to document and appreciate the extensive work that is already being carried out in higher education to address the climate crisis, to assess its effectiveness in practice and to identify gaps, as well as providing a basis for planning and future work.

## Five modalities of the university

University activities can be divided in different ways, the most common being the triad of teaching, research and community engagement. Scholars of higher education have, at times, broken these activities down into a larger array of functions. In relation to climate change specifically, Henderson et al. (2017) based their analysis around five domains: governance (institutional priorities, values and proclamations), education, campus operations, research and community outreach. Findler et al. (2019a), focusing on the broader area of sustainable development, identified five similar areas: education, research, outreach, campus operations and campus experiences, in addition to an integrative impact of the HEI as a whole. These schemes do not present major divergences but highlight specific areas for emphasis. For the purposes of this book, the university will be understood as having five modalities of action, as previously outlined in McCowan (2019): education, knowledge production, services, public debate and campus operations. Here, the broader learning acquired by students that is designated by Findler et al. as 'campus experiences' is included under 'education', while governance (outlined in Henderson et al. 2017) is not considered a separate domain, as it is understood to be underpinning and expressing itself in relation to all of the areas.

The five modalities employed in this book are outlined in Figure 4.1, along with some examples of activity in each. The first two modalities – *education* and *knowledge production* – correspond to the most recognisable 'pillars' of the university: teaching and research. The first, *education*, refers to the role of the university as a space for learning, and for personal, civic and professional development. It is the most prominent function of the university, and many HEIs only have this function. *Knowledge production*, on the other hand, involves not the transmission or facilitation of knowledge, but its generation, and normally arises from research and scholarship carried out by academic staff, but in some instances also by students and community members. This modality includes not only basic and blue skies research, but also knowledge applied to the practical demands of government, industry and civil society organisations, the development of new forms of technology and innovation more broadly.

In conventional categorisations, the third pillar of the university (in addition to teaching and research) is the least well defined, and is variously known as service, community engagement, extension or third-stream activity. It refers to those activities of the university that connect directly with external communities – that is, not with their own staff or students. Here, these activities will be divided into two as they constitute very different types of work: provision of services and public debate. In relation to the former, there are services delivered directly to communities – for example, running a health or legal clinic that community members can access, monitoring levels of air pollution to provide information on when it is unsafe to go out, or offering a short course on business French. This category also includes services provided to government, organisations and business, such as consultancy and secondments.

Yet there is a broader set of public engagement activities that relate to debates in the public sphere, taking place through the ideas put forward in formal research outputs, such as journal articles, which filter their way through the media into public discussion, or through the direct engagement of staff in the media or social media. In some cases, universities will have their own media outlets, such as newsletters, blogs and radio or even television stations. This modality can also express itself through the political involvement of staff and students, their participation in campaigns and protests, and in other forms of direct action. Universities can also serve as sites (either physical or virtual) for hosting and encouraging deliberation and debate, as discussed by Marginson (2011) in relation to the 'public sphere' mode of the public good. The level of influence on public debate differs markedly between universities, between countries and from epoque to epoque, and is very hard to gauge, but undoubtedly represents a significant influence of the institution on society.[1]

The use of the term 'services' here might raise some eyebrows, given the associations of service delivery with neoliberalism, and its apparently monodirectional dynamic – sending something from university to community. Most in the field prefer an expression like 'community engagement', which allows for multiple and diverse forms of (bidirectional) interaction. As explored further in Chapter 9, this book also endorses these forms of interaction in a normative sense. However, for an analytical scheme, the term community engagement lacks precision. What goes under the umbrella of 'community engagement' could variously be *education* (for example, a series of workshops on forest regeneration organised for community participants), *research* (a collaboration between university staff and community members to collect data on growth rates of newly planted saplings), *service provision* (the building of a greenhouse for keeping seedlings) or *public debate* (joint construction of a manifesto on reforestation for influencing local policy). While the partners and relationships may be similar, the kinds of activity and their implications are very different. This scheme in fact focuses on one part of the puzzle – what the university does – rather than the actions of its external partners and interactions with them: the latter are, of course, important but can only be adequately covered by other frameworks.

There is a sphere of university activity that does not represent the core purpose of the institution, but is nevertheless very relevant to climate change, and that is its operation as an institution and as a campus. As a community and as an organisation, the university manages its finances and its human resources – purchasing equipment, using fuel, selling food and merchandise, and in some cases making investments – and all these activities have implications in terms of mitigation and adaptation of climate change. In some cases, universities own land beyond their immediate campuses, and make decisions about the usage of that land, for agriculture, forestry or commercial developments. The fifth modality, therefore, covers this area of campus operations. In this category, we might also include the travel undertaken by international students, a significant source of carbon emissions: while these activities may often be categorised within the 'education' category, it is not strictly a result of the teaching and learning itself, but of the logistical organisation of the institution and its members. For some institutions the goal in terms of campus operations is to become carbon neutral or net zero – which can involve not only reducing emissions, but also offsetting through carbon credits or sequestering carbon. The five modalities are represented in Figure 4.1.

These five modalities are not quite exhaustive of all of the activities of the university. As outlined in the discussion of 'functions' in McCowan (2019), there are also the activities of archiving and validation. The former involving libraries, collections and information services has certainly become less important in the internet age, with a move from physical to digital storage of information and the consequent ease of distribution, yet universities still retain an important role in relation to specialist collections, physical artefacts and, more broadly, in curation. Validation is still a highly important function in terms of certification of degrees (individual knowledge) and peer review in academic journals (collective knowledge). We could understand the individual and collective knowledge validation functions as being part of education and knowledge production modalities respectively. However, it must be recognised that there are elements of science and scholarship, and activities of the academic community more broadly, that cannot be fully captured in this scheme.

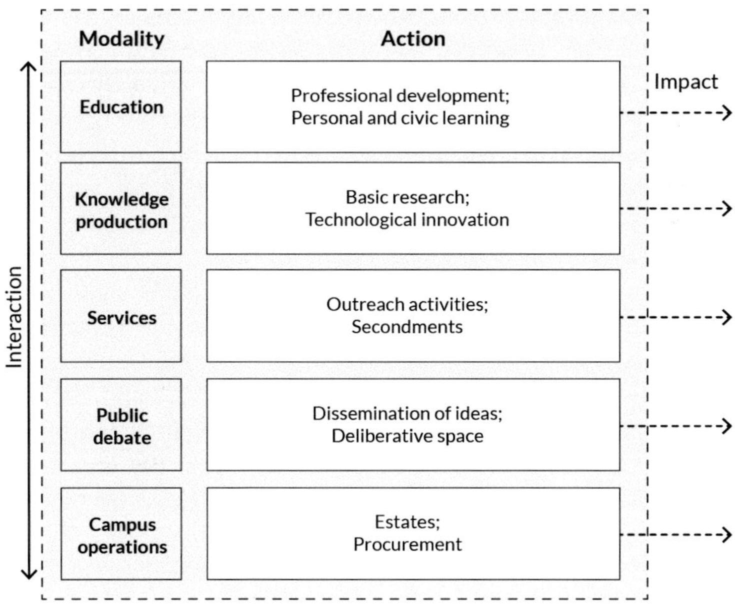


**Figure 4.1**: University modalities.
Source: The author.

For each of these five modalities, it is important to observe three characteristics: the *action* itself, and the nature and extent of activity undertaken within each modality; the *interaction* between them; and the *impact* they have on the society outside.[2] The most straightforward of these characteristics is the first. We can identify the extent to which institutions carry out particular actions: for example, an institution may be involved predominantly in education, or it may be a research organisation, but with graduate-level courses. Some universities have extensive work in the area of delivery of services to community and public engagement. The nature of these activities also differs markedly. There will be different focal points, involving various actors and underpinned by diverse values. Crucial here is the extent to which the modalities are oriented towards public and private good (Marginson 2011, 2018; Singh 2012; Unterhalter et al. 2017), and, in relation to climate change specifically, the extent to which those actions are relevant to adaptation to and mitigation of climate change.

Can we determine that certain of those five modalities are the 'core' business of the institution and others more peripheral? That may be possible if it can be shown that some are dependent on the existence of others. Services and public debate for the most part depend on education and knowledge production as they are an application, extension or broader communication of those activities. Campus operations constitute the organisation of the other activities so are certainly dependent on them. Do education and knowledge production rely on each other? While they can be strongly interrelated, it is possible for them to exist independently: there are research institutes that only have this function, and it is also possible only to be an education institution, teaching the broader store of humanity's knowledge, rather than the knowledge it has itself created.

Another criterion determining core status might be those activities that can only – or most effectively – be carried out by the university. Education and knowledge production would seem to fit this category, particularly when they are intertwined. On the other hand, universities may offer valuable provision of hospitals and legal clinics, but these services could be delivered by other institutions. For both of these reasons, therefore, both education and knowledge production would appear to be the 'core' activities of the institution and the other three emerging from them (on Figure 4.1, for ease of visualisation, they are side by side but a layered representation may be more reflective of this dependency). Nevertheless, each of the five has its own independent dynamics and emergent characteristics. New knowledge can be created through public debate and community work, in ways that might depart from the basic research developed in the university and be unanticipated by those involved.

Regardless of their analytical relationship, in practice there do exist hierarchies and power disparities between these different areas. Research has pride of place in the contemporary elite university, bolstered by the weighting given to it in international university rankings. Community engagement in its different forms usually occupies a subordinate position, either with a posture of beneficence towards the less enlightened external world that finds space when academics have time, or with a commercial orientation that aims to generate income for the institution. Some more subversive approaches may attempt to revert that imbalance, as with, for example, Santos's (2004) ideas of counter-extension, in which the university draws on and learns from the knowledge from the community. In some cases, the university may be in a subordinate position to the external 'community', when the external actor in question is a government or powerful corporation, and is the university's paymaster.

It is also important to consider the interactions between the five modalities. An institution may have greater or lesser porosity between these different activities (McCowan 2019). Just as with the curriculum and disciplinary distinctions – analysed by Basil Bernstein (1971) in terms of weak and strong classification and framing – the modalities vary from institution to institution in the extent to which they are merged or held separate, and whether there is movement of people and ideas between them. Most commonly discussed is the teaching–research nexus, the extent to which lecturers incorporate their research findings into taught courses, or alternatively the opportunities available for students to participate directly in and benefit from research projects. Community engagement work, either of the service delivery or public debate type, can also draw to a greater or lesser extent on research and scholarship carried out within the university. The fifth modality of campus operations will necessarily have a very high level of interaction with the others, since by its very nature it underpins all of the core functions of the university.

There may also be different types of interaction – it may be a simple sharing or diffusion of material from one to the other. So, for example, a new technique for low-cost water purification developed within the university may be rolled out to surrounding slum areas. Yet, in other cases, the value of the interaction may be greater than the constituent parties: so, it may only have become possible to develop the low-cost water purifier through the engagement work between scientists and community members, drawing on the experience of both (in other words, *co-production*).

There are different ways in which interaction between the modalities can manifest itself and be achieved. First, the modalities can be aligned through the design of university leadership or strategy. For example, structures or incentives might be established so that taught courses, research agendas and community engagement programmes all adhere to an institutional principle (take, for example, the five grand challenges of my own institution, UCL: climate crisis, data-empowered societies, inequalities, mental health and well-being, and intercultural communication). In this case, there is not necessarily any direct interaction between the modalities, but they are brought into line with an external principle or edict. Second, there may be a flow of influence from one or more of the modalities to the others. So, for example, a research-intensive institution may show strong uptake of the ideas generated through its research programme in all of its other activities. The third type of interaction – cross-fertilisation – also shows direct influence but, instead of a one-directional flow, involves actual encounters between people and between ideas, and joint change. An instance of this is provided in the Climate-U project (Sasidevan and Santha 2023): as part of the Master's in Livelihoods and Social Work, the Tata Institute of Social Sciences in Mumbai has incorporated principles of participatory action research into the programmes for all their students. Each student spends a period in a rural village working with villagers on an environmental and social development project, then writing it up for their dissertation: thereby integrating elements of teaching and learning, research and community engagement.

We can therefore observe four kinds of relationship that the modalities can have on each other: (i) *separation* – little or no interaction; (ii) *alignment* – top-down steering to ensure that all follow the same principles; (iii) *influence*: one or more of the modalities determining the content of the others; and (iv) *cross-fertilisation*: contact of people and ideas leading to mixing and synergies. From a normative perspective, there is clearly much to be lost from low levels of porosity between the modalities. Inert and out-of-date curricula result from isolation from research, while the possibilities of research are limited without engagement with external communities, and students miss out on learning opportunities without engagement in their campuses and beyond. The value of these interactions between modalities is brought out strongly in Dilly Fung's (2017) *Connected Curriculum*, for example. We can start to see the possibilities and perhaps the need for intermodality and transmodality work – just as we can interdisciplinary and transdisciplinary curriculum and research.

Finally, there is impact. This element concerns the outcomes of a particular activity: for example, the effect of an undergraduate course on the life and work of a graduate, or the changes in society resulting from research in biochemistry or anthropology. There is a high degree of complexity in gauging impact, in the first place due to challenges of tracking and attribution (McCowan 2018). We would expect history graduates to have developed a set of positive values, knowledge and skills through their studies, and that those qualities would positively influence their lives, work and interactions with others subsequently. Yet it would be almost impossible to fully gauge the millions of interactions they will have with others through the course of their lives. Second, it is not easy to attribute changes that are observed in the outer society directly to the influence of the university. While history graduates may be using documentary analysis and critical thinking skills in their environmental work as civil servants, it is hard to say what is down to the impact of the university, and what the impact of their previous schooling, spare-time reading, interactions with family and so forth.

So, the impact of the university in these diverse areas is extremely hard to gauge with any precision (McCowan 2022). Nevertheless, it is important that we attempt to do so, while acknowledging that any endeavour of this sort will be an approximation. The following section outlines an attempt to understand this dimension in greater detail, and outline the pathways through which the university can have impact, relating them specifically to climate change.

## A framework of university impact on climate change

The visual model (Figure 4.2) representing the trajectories of impact contains four stages, starting with the university itself, divided into the five modalities outlined in the previous section. It shows the general movement of impact of the university on society and the natural environment from left to right across the diagram, as well as the feedback loops from right to left, indicating the effects of the environment on society, and of society on the university.

The stage of 'university' involves primarily the actions of its members – staff and students – but also those of the institution itself, in terms of its organisational structures, carbon emissions and investments. After

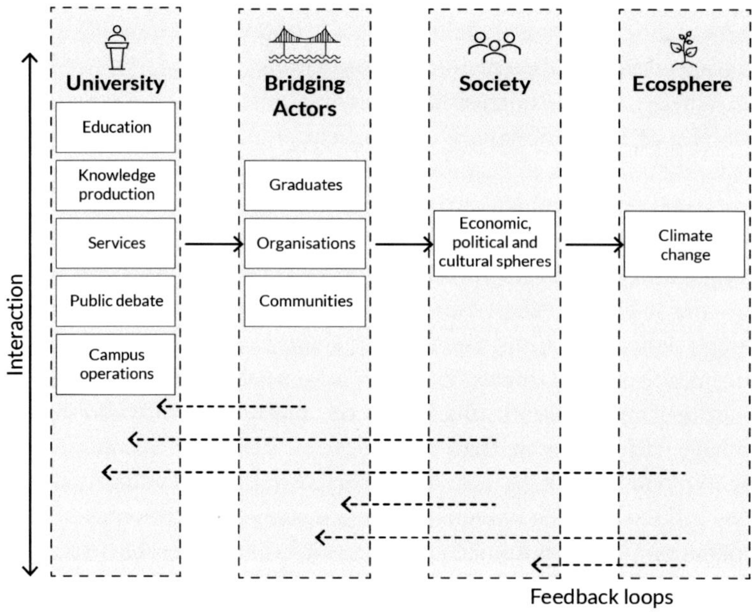


**Figure 4.2**: Stages of impact of the university on climate change.
Source: The author.

the university modalities, come 'bridging actors': the groups outside the university that have direct interaction with it. Most obvious of these, and by far the most numerous, are the students, who having left the university go out into the world as graduates. Universities also have direct contact with external communities, including businesses, government and members of the local community, through their research and community engagement work. Universities deliver services to these organisations through consultancy work, running projects, industrial collaboration, writing reports, seconding staff members and so forth. Some organisations also commission research from universities – although this is normally limited to businesses, foundations, large non-governmental organisations (NGOs) and government departments. These actors are designated 'bridges' since, in addition to receiving impact themselves, they also serve as conduits of impact to the broader society.

The next stage is that of society as a whole. In this case, it is not a question of contact between the university and specific people, groups or organisations. Instead, it is a more diffuse interaction of ideas, products and influences, one that is harder to chart and attribute. So, for example, the university may develop a vaccine that is adopted for general use, or achieve a breakthrough in mathematics that influences a new generation of computers used at home and in the workplace. In this case, the impact goes straight from knowledge production to society, highlighting that these benefits occur, even when individuals have not been directly involved in the process or commissioning of research. In other cases, it may be a question of knock-on impact, going via the bridging actors stage – so, for example, when others are subsequently influenced by the work of professionals who have been trained in universities. (These different trajectories are outlined in Figure 4.3.)

What this means, therefore, is that universities, through the education modality, influence society as a whole, even those members who have not attended university. This influence occurs through the knock-on effects of the learning acquired by graduates, primarily through their employment, but also through their civic participation and personal lives. So, all people benefit from the teaching and learning taking place in university through the subsequent work of doctors, engineers and social workers, for example. There is extensive empirical evidence (for example, McMahon 2009; Bynner et al. 2003; Oketch et al. 2014) showing the general impacts of graduates on society in the areas of stronger support for democracy, human rights and environmental protection, lower crime rates, gender equality, and better nutrition and health.

Universities also have an impact on society through public engagement activities. In some cases, these activities may be closely related to knowledge production: for example, media discussions of research that has been carried out, or popular books and television programmes on science – of the kind produced by Richard Dawkins and Brian Cox in the UK. In other cases, they may not be directly connected with research carried out in the university and may involve political opinion, expressed through social media or formal media channels, or other commentary on society, with some academics taking on the 'public intellectual' role.

The influences of the university on society at this stage may involve concrete changes in the lives of individuals (protection against an infectious disease) or shifts in their thinking (understanding the impact of livestock farming on greenhouse gas emissions). Yet it may also contribute in a more diffuse way to the constitution of economic, political and cultural structures in society, influencing norms and social practices, as well as policies and institutions.

Finally, there is the fourth stage, the 'ecosphere'. Here the emphasis moves from human societies to the natural world, and the influence that the former has on the latter. For the most part, the influence is mediated by the communities that have direct contact with universities, and the knock-on impacts on society, that is, via stages two and three. In some cases, there are direct impacts – for example, through campus operations, with the effects of energy usage, recycling, procurement policy and so forth. It is important to point out that the impacts at this final stage are normally only observable in the long term, and as part of the general human influence, rather than being easily isolated in the short term.

This framework focuses mainly on the flow of movement from left to right, but the inclusion of the feedback loops shows that there are, in reality, flows of different types in both directions. The natural environment can have significant impacts on society and universities, including destructive ones, and there are also impacts from society and its different actors on the university. While Figure 4.2 acknowledges these potential effects (ones that make necessary the adaptation pathways), the main focus in this book is on the impacts of higher education on climate.

## Mitigation and adaptation pathways

The following are some possible trajectories in practice of the positive influence of the university on climate change, as shown in Figure 4.3 below. A total of 15 pathways are outlined, 10 of mitigation (labelled with M) and 5 of adaptation (labelled with A). These do not quite represent an exhaustive list – there are other conceivable routes – but they do constitute the most important and most commonly observed ones.

This figure has only assessed pathways for mitigation and adaptation, but pathways could also exist for regeneration – following similar trajectories of interaction between the university and external elements, but with some divergences in activities and outcomes focused more on the revitalisation of ecosystems for long-term sustainable living. The key in Table 4.1 provides an additional explanation of the pathways and the actions and impacts occurring at each stage.

As is evident in Figure 4.3, all of the pathways start with one of the five modalities, but do not follow a uniform course, and do not necessarily manifest at each of the four stages. First, adaptation measures do not aim to bring about an impact on the ecosphere, so they do not 'reach' the final stage. And second, for mitigation measures, some of these occur within the university, some in specific individuals, communities or institutions, while some require widespread societal uptake. In some cases, the influence goes directly from university to ecosphere – for example, in the case of pollution emitted by university buildings. Initiatives relating to mitigation, by definition, involve the impact continuing all the way through to the ecosphere.

There are a larger number of pathways relating to mitigation, not because of its greater importance, but because of the complexity of its trajectories. While the pathways are identified as being either mitigation or adaptation, it might be possible for actions taken to constitute both at the same time. In practice, many actions that are adapting to climate change may also be mitigating it – for example, tree planting can reduce risk of flooding and also reduce levels of carbon

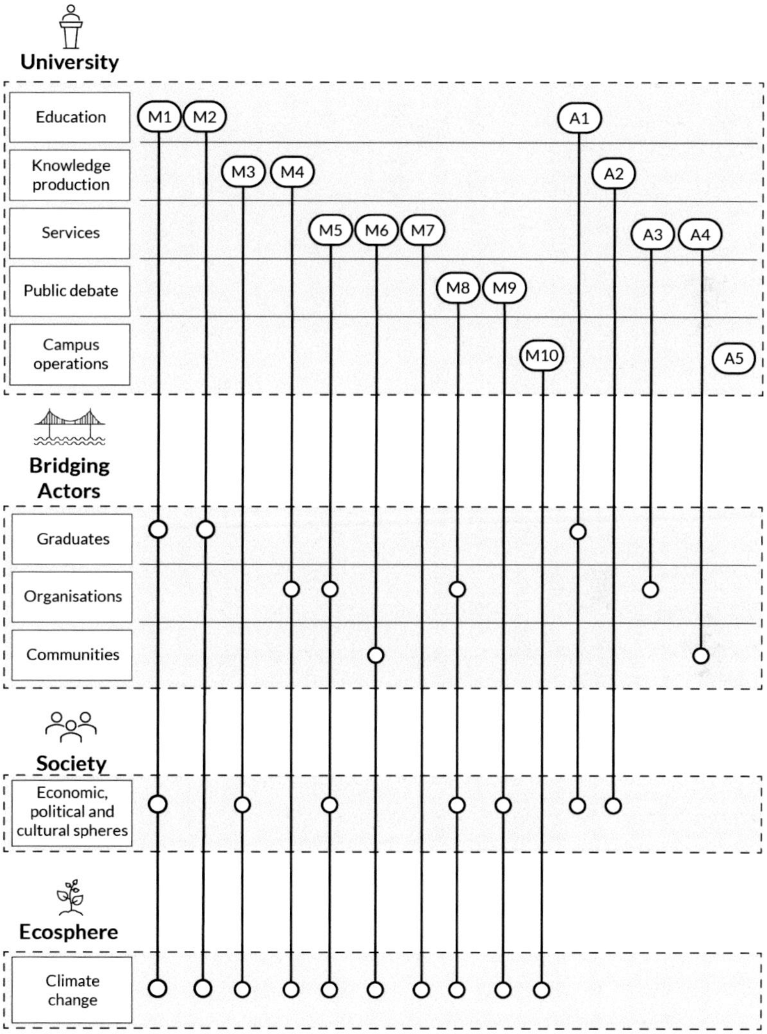


**Figure 4.3**: Mitigation and adaptation pathways for university influence on climate change.
Source: The author.

**Table 4.1**: Key to mitigation and adaptation pathways for university influence on climate change.

| Mitigation pathways | |
|---|---|
| M1 | Student acquires professional knowledge relating to climate change at university; applies that knowledge in the workplace and has an influence on others; impact on societal causes of climate change; impact on climate change mitigation. |
| M2 | Student engages in broader learning and personal and civic development while at university; influences thought and action in later life; impact on climate change mitigation. |
| M3 | Research is carried out on the climate; enhances societal understanding of the changes under way, their causes and possible responses; impact on climate change mitigation. |
| M4 | University develops new products or technology for addressing climate change; taken up by government, business or civil society organisation; impact on climate change mitigation. |
| M5 | University provides a service directly to an institution; institution develops policy; policy influences actions and behaviours; impact on climate change mitigation. |
| M6 | University provides a service directly to an institution; institution enacts environmentally friendly measures; impact on climate change mitigation. |
| M7 | University provides direct environmental service; impact on climate change mitigation. |
| M8 | University engages in advocacy, campaigning or mobilisation; influence on government policy; policy influences societal actions and behaviours; impact on climate change mitigation. |
| M9 | University engages directly with the public in relation to climate change; change of understanding of climate change occurs, leading to some change in consumer and corporate behaviour; impact on climate change mitigation. |
| M10 | University alters its own institutional functioning; impact on climate change mitigation. |

| Adaptation pathways | |
|---|---|
| A1 | University develops knowledge, skills and values in students; graduates have enhanced ability to adapt their lives to changing climate; knock-on impacts on broader society. |
| A2 | University develops new products or technology for adaptation; adopted by population to address impacts of climate change. |
| A3 | University provides advisory services to company; company adapts its operations in light of climate change. |
| A4 | University provides service to communities; communities enhance their ability to adapt to changing climate. |
| A5 | University adapts its own functioning to the changes brought by climate change. |

**Table 4.2**: Positive impacts of universities on climate change.

| Pathway | Area of activity | Example |
|---|---|---|
| M1 | Professional development | An engineer applies principles of sustainability in her building designs. |
| M2 | Personal transformation | A graduate has acquired basic knowledge of impact of greenhouse gases on the climate, and moves towards using renewable energy sources in his own house and transport. |
| M3 | Research and scholarship | Paleoclimatological reconstruction of temperature rises over the past 1,000 years enables an understanding of the extent to which global warming is anthropogenic. |
| M4 | Application of knowledge/ innovation | Geo-engineering technology is developed for blocking the sun's rays. |
| M5 | Secondment | An environmental scientist is seconded to a government department to lead strategy on climate change. |
| M6 | Community engagement | A university works with a housing association to make their energy usage more efficient and reduce fossil fuel emissions. |
| M7 | Environmental service | Students organise large-scale reforestation programme in areas surrounding the university. |
| M8 | Campaigning and mobilisation | University lecturers write open letter denouncing government subsidies to fossil fuel companies. |
| M9 | Awareness-raising | A university lecturer runs a television series discussing how individuals can change their lifestyles to be more environmentally friendly. |
| M10 | Campus sustainability | The divestment of university endowment from shares in fossil fuel companies takes place. |

| Pathway | Area of activity | Example |
|---|---|---|
| A1 | Personal transformation | Graduates have access to the latest research and advice around flood risks and adapt family homes accordingly. |
| A2 | Research and scholarship | Researchers develop new technology for reducing soil salinity in coastal areas affected by rising sea levels. |
| A3 | Application of knowledge/ innovation | Research findings allow a business to identify a site for relocation to avoid rising sea levels. |
| A4 | Community engagement | A university provides a training course for local farmers in developing new crops that are appropriate for changing weather conditions. |
| A5 | Campus sustainability | A university building is relocated, to avoid site vulnerable to mudslides in heavy rains. |

**Table 4.3**: Negative impacts of universities on climate change.

| Pathway | Example |
|---|---|
| M1 | Students develop professional competencies that allow them to increase fossil fuel extraction. |
| M2 | Students acquire attitudes privileging maximisation of profit over protection of natural environment. |
| M3 | Scientists funded by fossil fuel companies produce research that casts doubt on existing climate science and undermines efforts at reducing carbon usage. |
| M4 | University develops new products and technologies that are dependent on fossil fuel usage. |
| M5 | Lecturers seconded to think tanks work to undermine environmentalist agenda. |
| M6 | A community engagement project encourages a local community income-generation scheme that causes local environmental destruction. |
| M7 | Students cut down an area of old-growth forest as part of a cash crop scheme. |
| M8 | Anti-environmentalist public intellectuals provide academic backing for regressive policies of populist government. |
| M9 | Climate denial book written by university lecturer encourages scepticism among public. |
| M10 | Universities develop new student accommodation buildings without environmentally friendly specifications. |

dioxide in the atmosphere – but the pathways are kept separate here for analytical purposes. Curricular interventions will very often deal with both these questions simultaneously, and buildings can be designed to be at once more resilient to extreme weather and more efficient with energy. Some examples of impacts for each of the pathways are outlined in Table 4.2.

The pathways outlined above, and examples accompanying them, assume that these influences will be positive. However, it must be acknowledged that negative influences are possible, and do actually occur in a number of cases. Universities emit greenhouse gases – directly through their own campuses, as well as through the travel undertaken by their students and staff, and in some cases through their investments in fossil fuel corporations (Grady-Benson and Sarathy 2016). Public engagement can have a negative impact in cases where professional scientists, or those with scientific training working for lobbying organisations or partisan think tanks, act deliberately to obscure public understanding of science or to propagate mistruths. Examples of this type are detailed in relation to the tobacco industry and global warming in the book and documentary *Merchants of Doubt* (Oreskes and Conway 2010). Some potential negative impacts for each of the mitigation[3] pathways are outlined in Table 4.3.

In some cases, therefore, universities need to develop new lines of work that can have a positive impact on climate change; in many cases, however, it is a question of turning around existing negative impacts, or neutral influence, into positive influence. For example, university-based engineers may continue to contribute to infrastructure development, but utilising net zero technologies. The notion of embodiment (McCowan 2019), therefore, becomes relevant here. One of the ways in which this framework can be used is to allow universities to assess the alignment of the work they are doing in these different areas with their overarching aims – to determine whether a mission to support the SDGs, for example, is being supported in all the different aspects of the work undertaken by the institution, or whether there are positive and negative influences running through different pathways.

## Characteristics of the model

The relationship between higher education and climate change shows many of the characteristics of complex systems, as outlined by Tikly (2019) and discussed in Chapter 2. Complexity is a characteristic of the climate system generally, of the higher education system, and of the interaction between the two. As discussed above, the system has multiple positive and negative feedback loops, and, being an open system, what happens inside and outside the university constantly modify one another. There are some elements of *autopoiesis* here, although it is not necessarily a self-regulating system in the sense of maintaining equilibrium, and there are many instances of instability.

Importantly, the system has the quality of 'emergence' – new action or being that evolves from the interaction between the elements, and which is not present in the original components. This quality is particularly crucial given the nature of the university as an institution focused on open-ended enquiry and the quest for human understanding (Collini 2012). Education and knowledge production have particular characteristics in that they involve reflection, enquiry and creation of ideas, and as such can modify their own nature in the course of the process. To give an example, a student may start out with an intention to pursue a course in business studies so as to become a successful entrepreneur, but through the process of reflection engaged in during the course decides to abandon this life course and become a Greenpeace activist instead.

At first sight, Figure 4.2 may give the impression of being a closed system, but only because it cannot represent all the external factors that can influence the processes. Crucial among these are the dynamics of political economy that support or constrain change, constituting the 'conditions of possibility' discussed by Unterhalter et al. (2017). Clearly, the work of the university does not emerge from nothing, and a complete understanding of the dynamics involved would include the constitution of the different modalities – why do education, knowledge production and so forth appear in the way they do? These are highly complex questions and involve a combination of immediate and direct factors, such as higher education policies and resourcing for the sector, and the preparation provided for students at lower levels of the education system, as well as deeper historical factors such as models of university and epistemic traditions. The purpose of this analysis is not to provide an account of the roots of university practice, so this part of the system has been left implicit (some discussion of these factors at the global level will be provided in Chapter 5). Nevertheless, through the feedback loops coming back from community, society and ecosphere, the framework acknowledges that there are extensive impacts back on the universities themselves. Furthermore, the arrow to the left of the university column (Figures 4.1 and 4.2) shows the interrelationships between the modalities and the influences that they have on each other.

Like all representations of human systems and dynamics, the neat separations between different stages and actors do not quite capture the reality: for example, people who are in the 'graduates' box (Figure 4.2) are simultaneously community members and work for governments, businesses and NGOs. Graduates act as agents of knock-on impact on others in society (for example, through their work as lawyers or computer engineers), but are also influenced by the work of other graduates in turn, and directly by the actions of the university. So the same people can be in the 'bridging actor' stage as in the 'society' stage, but fulfilling a different role in each case. Nevertheless, it is useful to separate out these categories in order to understand how these people mediate the influences of universities in different ways depending on the roles they are playing.

Figure 4.2, of course, does not attempt to portray the entirety of the learning system – formal, non-formal and informal – and all the processes of knowledge production in a society, a task which would barely be possible in a single representation. The attempt here is only to look at how universities interact with climate change. In addition, it cannot capture all the purposes and actions of universities, and their interactions with different spheres of society, but focuses only on those of relevance to climate change. Even in relation to climate, the interactions are more extensive than those represented through the pathways. For example, community members are involved in knowledge production through participating in focus groups, clinical trials and so forth.

In an earlier study of impact (McCowan 2018), six dimensions of impact were identified: source, form, trajectory, intensity, timescale and destination. Of these, source, trajectory and destination are represented in Figure 4.2, indicating the 'origin' of the impact (say, a research project), its 'trajectory' (uptake of a new development by a solar panel company) and its 'destination' (public administration buildings seeking to go carbon neutral). The usual caveats are necessary here around isolating causes and effects: these impacts do not originate purely in the university, they do not move on an entirely linear course, and they may have diverse destinations of impact – so the pathways charted are approximations rather than absolute categories.

However, we also need to take into account the other three dimensions of impact: form, intensity and timescale. As regards form, the flows along the arrows in Figure 4.2 involve ideas and actors. While the two cannot exist separately, there is some distinctiveness of each: an output of the university might be a research paper – take, for example, Mann et al.'s (1998) seminal article on changes in climate, which led to the popularisation of the 'hockey stick' graph. The article is, of course, created by human beings, and is subsequently used by them, but the output itself is not in the form of a person, and is not confined to a particular person or set of people in its subsequent trajectory. In observing the impact of universities through ideas, we can distinguish between economic, political and cultural spheres – which, while all having a knock-on effect on climate change, will do so in different ways.

On the other hand, the output of the university might be in the form of persons, say a graduate in marine biology, who then goes on to work in the field of conservation of ocean life, and make a positive impact in that area. The graduate has acquired ideas within the university, and is employing them in her work, but it is not a specific idea or set of ideas that is making the impact, but the human being who combines them in particular ways in response to a specific set of external circumstances and problems to solve. In addition to ideas and actors, some outputs of the university are material products – for example, more affordable solar panels or a new form of combustion engine used in aeroplanes to lower emissions.

The intensity of these different forms of impact varies greatly, depending on the 'force' of the original intervention by the university and the resources deployed to maximise its impact subsequently. There are also variations of intensity depending on the stages through which the pathway passes, and the breadth of its reach – having potentially a strong impact on a small group of people, or a more diffuse influence on a large group or on humanity as a whole. Universities emit carbon dioxide, which has a direct impact on the atmospheric conditions affecting the climate, yet the emission constitutes a small proportion of all the carbon dioxide in the air. On the other hand, the education provided by universities may have a profound impact on the life of an individual, enabling further knock-on effects – perhaps enabling her to obtain a job in a local council, through which she is involved in establishing stricter regulatory codes to reduce the number of high-polluting vehicles. Any single action by a university will, of course, have a very small impact on temperature rises, although taken together the effects may be substantial.

Lastly, there is timescale. There is a time lapse between the different stages – between those actors who come into direct contact with the university, the broader society and then onto the ecosphere. In the framework put forward by Findler et al. (2019a), there are two stages of impact – direct and indirect; the former evident in the short term (for example, use of research findings by a corporation), and the latter in the longer term (changing business practices). Direct and indirect impacts are indicated on the framework presented here, but should not necessarily be identified with the different stages of 'bridging actors' (direct) and 'society' (indirect): in some cases, there may be direct impacts on society as a whole (development of new electric car technology), and even some direct impacts on the ecosphere (reduction in greenhouse gas emissions). Direct/indirect should also be distinguished from short/long term, although they may often coincide.

Nevertheless, there are significant differences of timing in the different forms of impact, which must be borne in mind given the urgent and time-bound nature of climate change itself, as well as the pressures on universities to show immediate tangible results of their work to justify their funding. The timescale element may interact with the other dimensions of intensity and form: the type of impact in question may change over time, or it might vary in its strength, either building up with time or dissipating.

A final point is that none of the pathways represented in Figure 4.3 are inevitable, and there is a degree of unpredictability in all the trajectories of impact. First, human agency ensures that the processes and outcomes of teaching and research are inherently unpredictable: while a lecturer may have a particular learning outcome in mind, and a researcher a question to be answered, the process of pedagogical interaction or enquiry may end up leading to quite a different end point. And second, the complexity of the climate system ensures that even well-thought-out intentions may have unexpected environmental consequences. This element of uncertainty does not require us to abandon all modelling of the processes, but to let go of any rigidity in our application to actual contexts.

## Final reflections

The framework outlined in this chapter is both analytical and normative. In an analytical sense, it sheds light on what the university is doing, on the diverse pathways through which it has an impact on the society outside, and the likely effects on climate change. It puts forward a frame that highlights first the trajectory of impact (moving from one of the five modalities of the university to various bridging actors, to societies and to the ecosphere), and identifies form, intensity and timescale as key dimensions to be observed. It can serve, therefore, as a tool for researchers in locating the focus of their existing research onto the broader map of university action, and also in drawing attention to new elements of the process.

Taking a broader view of the literature as a whole, the framework can reveal the emphases and also the silences and gaps. Research to date has focused predominantly on just some of these pathways (M1, M2 and M10) – those relating to changes in the curriculum, and campus sustainability. There has been less attention to the knowledge production, public debate and service provision activities of the university, and as a result we have only a partial understanding of the role of universities globally in these areas.

From a normative perspective, universities can use this framework to ensure that action is being taken across the diverse spheres of activity of the institution, and to assess the extent to which potential synergies are being exploited. There has been a welcome increase in attention given to the environmental impact of university buildings, procurement and energy usage, along with efforts to measure and audit progress in these areas (Findler et al. 2019b; Vaughter et al. 2013). Yet the efforts of universities in other areas have been more limited (Henderson et al. 2017). While there has been an increase in taught courses relating to climate change, and some integration into natural sciences and engineering, the topic is only sporadically covered in other disciplinary areas (Vaughter et al. 2013). There is extensive research on the climate itself but much more work is needed in social sciences, arts and humanities to capture the deep societal roots of the question. Work in the areas of service delivery, public awareness and outreach (Hansen and Lehmann 2006) can be expanded significantly. Finally, a perennial challenge for universities is in ensuring that interactions between the modalities are maximised, synergies exploited – most obviously between teaching and research, but community engagement work and the campus too – and conflicts avoided. An ever-present contradiction, for example, is between the sustainability principles espoused by universities and their internationalisation strategies – usually involving extensive travel of students and staff with corresponding carbon emissions.

Clearly, these actions taking place within the university do not occur in a vacuum. As argued by Robinson-Pant (2020), the danger with systems thinking in education is to understand them as closed systems – the framework outlined above is subject to constant influences from the society outside, and also generates its own emergent dynamics. The actions within each of the university modalities are made possible by political, economic and cultural forces acting on the university, and are more immediately conditioned by prevailing higher education policies, at the current time dominated by a combination of marketisation, status competition through national and international rankings, and an incipient process of unbundling[4] (Marginson 2011; McCowan 2017, 2019). Higher education is also locked in a perennial tension between the production of private goods, which are disproportionately co-opted by privileged groups, and the production of public goods that can benefit all in society.

Chapter 5 will outline some of these global dynamics relating to the hierarchies and disparities of funding and prestige in the international higher education system, and assess how these pathways of impact on climate change manifest themselves when we look beyond the institutional to the systemic level.

## Notes

1 There are some areas of ambiguity here: massive open online courses (MOOCs) represent a liminal case, as they might be considered either education or public engagement. If the MOOC is part of a formal taught course, then it is more appropriate to consider it education, and if it is made available with open access for all without assessment, then public engagement. It is acknowledged that there are, in all cases, blurred lines between these categories.

2 Readers familiar with the earlier work *Higher Education for and beyond the Sustainable Development Goals* (McCowan 2019) may wonder what the relationship is between the scheme presented here and that of value, function and interaction. The framework presented here pays less attention to 'value', as it is assumed that the aim in question is climate action. Function maps closely onto the five modalities presented here, although with some differences of detail. Interaction relates to the movement (here termed impact) presented in Figure 4.2 across bridging actors, society and the ecosphere; the characteristics of that interaction will be discussed further in Chapters 5 and 9. In the earlier work there was no explicit reference to interaction within the university, between the modalities/functions.

3 For the adaptation pathways, what is observable for the most part is a lack of positive impact rather than any actively negative impact.

4 The disaggregation of the different functions and activities of the university into smaller units for the purpose of greater efficiency, personalisation and commercial gain (McCowan 2017).

# 5
# Global dynamics of higher education

The previous chapter has shown the ways in which a university can have an impact on climate change, through its diverse functions and interactions with society. However, for the most part the analysis has assumed an institution acting alone – at least, alone in relation to other HEIs. In practice, universities are part of systems, and are influenced by the norms and regulations of those systems, as well as by the other institutions within them, and their impact radiates out together with, and interacting with, the impact of the others. These systems are most obviously national: HEIs (with only a small number of exceptions, such as the UN University) are subject to the laws and regulations of national systems of higher education, which confer licences to operate, provide funding, regulate and audit the number, focus and quality of courses, and monitor outcomes. National structures do not determine altogether what happens in higher education, given the partial autonomy that universities have from the state, and the difficulty of regulating all parts of university practice. Yet they represent the conditions within which institutions can operate, and naturally have a strong influence on their functioning and therefore their impact on climate change. The nature of national systems, the role of the state (and market) and the level of autonomy of HEIs differs markedly from country to country, and awareness should always be maintained of these national and regional differences.

Nevertheless, institutions are not only subject to influence at the national level but also at an international level. This international influence has a number of conduits: first, the presence of globalised norms and ideologies of what a university is, how it should operate and what constitutes excellence (represented most prominently by the international university rankings); second, the existence of international organisations that propagate policies and practices of higher education and, in some cases impose them through conditional loans and other mechanisms; third, deliberate policies of internationalisation from the HEIs, which expose them to international influences and interactions; and fourth, the existence of scientific communities, disciplinary societies, and other coalitions and associations, both formal and informal, that go beyond national boundaries. Undoubtedly, there are also further reasons beyond those listed here.

A single book cannot hope to cover all these international and national dynamics in detail, and indeed they have been studied extensively in higher education research literature (for example, Knight 2022; Marginson 2024; Moscovitz and Sabzalieva 2023). This chapter will focus only on those dynamics of particular salience for impacts on climate change, and especially at the international level, given the global remit of the book. This chapter will have three sections: the first looking at the major trends influencing higher education at the global level, the second focusing on internationalisation, and the final one reflecting on the dynamics of impact on climate change in a global higher education system.

## Global influences

There are a number of mechanisms at the global level that influence the ways universities operate: the rise of international university rankings and the scramble for world-class university status, the dominance of neoliberal ideas of institutional financial self-sufficiency and entrepreneurialism, employability as the primary raison d'être for going to university, and a number of others. Ultimately, all of these trends influence either the resources that HEIs have at their disposal, and therefore their ability to operate and pursue their ends, or the ends themselves – their orientation, aims and modus operandi. In many cases, they may influence both the resources and the ends together, in different ways. Using imagery from sailing, these trends could be described as influencing, on the one hand, the strength of the wind in the sails and hence the speed of the boat, and, on the other hand, the course on which it is set. Addressing the climate crisis requires both of these things, with universities having both the resources (the wind strength) and the appropriate mission (the course set) to grapple the challenge head on.

### Wind in the sails

Early universities relied on some combination of patronage and student fees for their upkeep (Carpentier 2019; Perkin 2007). In the twentieth century, the marked growth of higher education systems, and their importance for nation-states in terms of ideology and nation-building, research for military and industrial might, and staffing for state bureaucracy led to increasing public support and control. The dominant model of higher education by the middle of the century (albeit for a restricted proportion of the population) was free-of-charge and state-run public institutions, with niches of philanthropic and religious private ones. The rise of neoliberal ideas from the 1980s, particularly in the Anglophone West, and the increasing strains on state finances through growth in enrolments, led to a dual process of retreat of the state: 'cost-sharing' in public institutions with the introduction of partial or total student fees, and the liberalisation of sectors to allow the entry of new private providers, including commercialised and for-profit institutions (Carpentier 2012; McCowan 2004; Salto and Levy 2021). Nevertheless, there is far from being a single story, and global regions and specific countries show distinctive mixes of public and private in both funding and governance.

These processes have led to an increase in overall resourcing for higher education, although with potential decreases in per capita funding, as well as significant disparities between institutions. (They also have marked implications for the orientation of higher education, as will be explored in the section that follows.) To give some examples of the resource disparities in today's world, annual government funding per student varies from under US$1,000 (Bangladesh) to over US$40,000 (Luxembourg), according to UNESCO Institute for Statistics (UIS 2023b) figures. Disparities exist not only between systems but also within them: in Brazil, fees for students in the private sector can be as much as US$2,000 a month for courses like medicine at elite institutions, while the average fee is US$200 and distance courses can be found for just US$70, with corresponding differences in quality of provision (SEMESP 2024).

Globalisation has intensified these disparities: developments in transport and communications allow students greater ease of mobility (either physical or virtual), while the upper middle classes have seen the tussle for valuable credentials transfer from the national to the global stage. Incoming international students have significantly increased the income of some universities (particularly in the fee-charging English-speaking countries, such as USA, UK, Canada, Australia), allowing them to invest in their infrastructure and research, and maintain a spiral of positive advantage. Compounding this disparity is the accompanying brain drain from low-income countries, as wealthier and high-performing students leave their countries to study and work in the Global North. For most institutions in the Global South, the vicious cycle of low levels of resources, poor infrastructure, poor conditions of work and brain drain has been insurmountable.

Not that all is plain sailing for universities in higher-income contexts. The intense competition for students in countries like the USA has led to an escalation of expenditure on campuses, leading to a spiralling of fees and costs that most commentators judge to be unsustainable – and possibly catastrophic (for example, Blumenstyk 2014; Goldrick-Rab 2016). For those institutions dependent on fees and international students, geopolitical shifts and changing preferences for destination country can be crippling.

Naturally, resources are not only a question of finance: what universities end up doing in practice depends on the energy and inspiration of those working there, not only on funding. Yet human resources (in terms of amount and qualifications) in today's world do to a large extent depend on financial resources, so – until such a time as we move towards a more frugal model of higher education – an adequate level of financing is essential.

Despite the retreat of the state, governments have attempted to counter these trends in order to ensure equity of access and also support research in the public benefit. Income-contingent loans, vouchers, fee waivers, score bonuses and a variety of other measures have been adopted in order to counter the severe inequities of access for lower-income students and other marginalised groups (Hoare and Johnston 2011; Meyer et al. 2013; Norões and Costa 2012). Most research funding is still public, and therefore able to support projects with public good benefits in areas such as climate science, although low-income and lower-middle-income countries usually lack substantial national research funding and rely on international collaborations.

At a global level, international and bilateral organisations have attempted to address these disparities through aid and cooperation funds. Data collected by the Organisation for Economic Cooperation and Development (OECD) in its Creditor Reporting System shows that aid to higher education has increased from US$1.8 billion in 2002 to US$5.3 billion in 2019, with the largest donors being: of the bilaterals, France and Germany; of the multilaterals, the European Union and the World Bank; and of the philanthropic donors, Mastercard Foundation (UNESCO IESALC 2022; Rensimer and McCowan 2023). These 'traditional' donors have now been joined by new actors – most importantly China, but also Brazil, India, Turkey and the United Arab Emirates

(Rensimer and McCowan 2023; Schendel et al. 2024). Nevertheless, over 70 per cent of these funds are allocated to scholarships[1] (mainly in the donor country or another high-income country) and therefore, while providing significant individual benefits for the recipients, they are often accompanied by the brain drain outlined above. Much of this support is also heavily enmeshed with geopolitics and knowledge diplomacy that is not always in the recipient country's interest. Some of this funding still opens important opportunities for universities in lower-income countries to engage in international partnerships; examples include the UK's Strategic Partnerships for Higher Education Innovation and Reform (SPHEIR) and the Norwegian Programme for Capacity Development in Higher Education and Research for Development (NORHED). In some cases, there have been more substantial investments in infrastructure – for example, the World Bank's US$60 million Nurturing Excellence in Higher Education Program in Nepal, although with problematic outcomes in terms of conditionality and homogenisation, as will be explored in the section that follows.

Despite these initiatives on the part of national governments and international organisations, higher education globally is still characterised by significant disparities of resources that represent both absolute inadequacy for many institutions around the world, and also relative disadvantage, both of which in different ways have an impact on institutions' ability to address the climate crisis.

## Setting the course

The global higher education system in the contemporary era presents something of a contradiction: on the one hand, it fosters sameness – a certain homogeneity of form and orientation – while on the other hand, it differentiates on the basis of prestige and resources, as seen in the previous section. In fact, this tension could be a positive one if it were reversed – with a diversity of mission and type, and an even level of quality and recognition (as argued in McCowan 2016). But in its current state it serves neither to provide equality of opportunity nor the kind of diversity of approach that might fully allow us to address the climate crisis.

For fear of overstating the case on homogeneity, it is important to acknowledge that there are some differences between universities in different parts of the world. World regions and their major higher education languages (including Chinese, Russian, Arabic, English, French, Spanish and Portuguese) have some divergences of academic culture, bodies of literature and traditions of scholarship. Most countries internally have some differentiation between institutional types – academic universities, technical institutions or polytechnics and specialised colleges. There are important differences between teaching-only and research-intensive institutions, and between public and private (although in many systems the boundary between the latter binary is blurred). Nevertheless, underpinning this differentiation are some foundational principles that are common across all systems. These norms emerge from the model of the European university, through its medieval and then Humboldtian manifestations (although recalling that what we think of as the European tradition absorbed intellectual influences from around the world), putting in place particular ideas about the teaching and learning relationship, certification of learning, the organisation of the university, the nature of academic work, the relationship between university and society, and disciplinary divisions (McCowan 2019; 2021b). These foundational principles have been joined in recent decades by a dominant political economy model of subservience to the market, both in terms of institutional organisation and the aims of the university (employability).

International rankings are widely recognised as being both instrumental in and symptomatic of the hegemonic conception of the university (Hazelkorn 2011; Molebatsi and McCowan 2022). The major international rankings – Shanghai, Times Higher Education and QS – reward elite status as measured through publications in exclusive journals, number of citations, research income, quality of incoming and outgoing students, and reputation, applying a series of self-reinforcing metrics. National-level rankings (for example, US News & World Report, and the Times/Sunday Times in the UK) employ different methodologies, and sometimes have less emphasis on research, but do not dramatically depart from the norm. These rankings do measure some things that are of value in the university, in particular high-quality research. The point is that they do not cover everything that is of value. In particular, they are mostly blind to the public benefit of HEIs, being unable to distinguish between universities that through their teaching and research promote the interests of their own constituencies or bring benefit to the broader society – having no means of doing so in their gauges of teaching and research quality, and paying almost no attention to community engagement.

International rankings customarily list 1,000 to 2,500 of the estimated 90,000 HEIs in the world (MacGregor 2022) and, while their influence extends beyond those institutions directly competing, they are primarily a model for elite institutions. Arguably more influential than the model of the 'world-class university' is an idea that extends to all HEIs – the idea of employability. Higher education has always had some connection to the labour market, particularly through the preservation and transferral of the bodies of knowledge associated with the professions of law, medicine and the clergy. Yet its relationship has become significantly closer over the past half century, both in terms of the raison d'être of the institution – justifying its public funding through contribution to the economy, and appealing to its customer base through career opportunities – and in terms of the expectations of its taught courses: producing 'work-ready' graduates, adding ever more specific professionalising courses and fostering transferable employability competences (McCowan 2015b; Tomlinson 2012).

As discussed in the previous chapter, the preparation of young people for work is an important component of climate action, if those graduates can act in ethical, environmentally sensitive and transformative ways in their workplaces. So there is nothing necessarily inimical between connection to the workplace and climate action. The problem with the current employability paradigm is that it subordinates all else – including ethics, public benefit and environmental impact – to career success. The effectiveness of a university course is measured by its ability to insert an individual into highly remunerated and secure employment (as demonstrated in the UK's Teaching Excellence Framework).

The corollary of the employability model in institutional organisation is the move towards business models of governance, drawing on ideas of neoliberalism. As explored in a previous work on the SDGs (McCowan 2019), universities have, since the nineteenth century, adopted, to differing degrees, elements of the 'developmental' model, with four characteristics: service to society, non-academic impact, egalitarian orientation and application of knowledge. These ideas have enhanced the community engagement function of universities and are crucial to efforts to maximise the institutions' impact on climate change. Yet, while these intentions are sincere (even in elite institutions), to a great extent they are overridden by the more pressing financial concerns brought about by the neoliberal model of institutional self-sufficiency. Dependent on student recruitment, corporate income generation and competitive research bidding, institutions can only perform their developmental function once they are safely in the black, or if they have a secure endowment. In contrast to the collaborative and coproductive forms of engagement that will be explored in Chapter 9, these models have encouraged extractive relations with communities, oriented around income generation and instrumentalisation.

Furthermore, there are an increasing number of for-profit institutions that, rather than mimicking businesses as many HEIs do, are *actually* businesses. These institutions now cover 9 per cent of enrolments in the USA (Hentschke et al. 2023) and more than half all new enrolments in Brazil (Pedrosa and Knobel 2023), with a significant presence elsewhere in Latin America and East Asia. In many cases, these institutions are part of large conglomerates, including some publicly traded companies. Customarily, they provide taught courses on a budget, no-frills basis, highly linked to employability, with no research or community engagement. Needless to say, the potential of these institutions for public benefit, and specifically climate action, is severely constrained.

Thus far this section has predominantly looked at negative trends in global higher education. There are certainly a number of worrying dynamics in the sector, but it is wrong to take for granted the struggles of previous generations, the achievements that have been made and the opportunities that still exist. The very existence of universities in our societies is a cause for celebration. And their significant growth over recent decades – while perhaps with dubious motivations – does provide at least the potential for transformative impact. Furthermore, while constrained by financial pressures and market ideology, universities nevertheless still maintain a commitment to transforming societies for the better – one strengthened by international solidarity and global networks of academics and students.

But global trends do not only operate from outside in, gathering pace in global societies and then sweeping universities away with them. Higher education is itself one of the motives for globalisation. The following section will look at the deliberate attempts of HEIs to internationalise themselves.

## Internationalisation of higher education

The history of higher education is one of international engagement – whether in the medieval European universities, Nalanda in India or the Islamic universities in North Africa – with student and staff mobility part of the fabric of their functioning. In addition, universities have rarely been tied to a national curriculum and have engaged with international bodies of knowledge and scholarship. To these characteristics have been added more deliberate attempts at internationalisation in recent years as a means of enhancing university quality, reputation, performance and finances.

Climate change is unavoidably a planetary question, as changes in temperatures and weather patterns are not confined to specific nations, and human actions in one location will affect those elsewhere. Responses to the climate crisis also require forms of international coordination. The engagement of universities with climate change, then, appears at first sight a 'match made in heaven'. Yet there are various factors upsetting this harmony. The international engagement of universities magnifies their negative as well as positive impacts, and there are a number of ways in which higher education is exacerbating the climate crisis. Specifically, there are forms of internationalisation of the university that have divergent and sometimes contradictory implications for climate change (Campbell et al. 2023; Rumbley 2020). Furthermore, developments in the sector through the twentieth and twenty-first centuries have challenged the internationalism of the early period of universities, with the prioritisation of national aims for higher education, and the emergence of locally focused institutions, leading to unevenness in international engagements (De Wit and Altbach 2021). These difficult questions make it crucial to explore deeply the complex relationship between internationalisation and climate action in higher education.

There is an emerging body of literature on internationalisation of higher education and climate change, with many professionals engaged in this area becoming increasingly aware of the sustainability implications of their work (CANIE 2022). Most of the publications to date have focused on the environmental impact of student mobility (for example, Nikula 2019; De Wit and Altbach 2020; Davies and Dunk 2015). Shields (2019), for example, showed that the carbon dioxide emissions from student flights were between 14.01 and 38.54 megatons in 2014, with the high estimate similar to the annual emissions of the entire country of Croatia or Jamaica. Some publications have also focused on the impact of academic staff travel for conferences and fieldwork (for example, Bjørkdahl et al. 2022; Gill 2021). Critical reflections on the impact of internationalisation have weighed up these emissions against the positive educational influence through intercultural understanding and global citizenship (for example, Ilieva and Tsiligiris 2021). Yet there is also a need to consider the broad range of activities of universities, including their research programmes, public service and campus operations, to understand their various influences and their relationships with each other.

The frame put forward in Chapter 4 represents a general tool for understanding the diverse trajectories of impact, to help us understand the various components of university work that we need to take into account, and to gauge the various configurations of interaction and impact of these elements. Yet how are these processes and outcomes influenced by internationalisation? Does the process of internationalising the university enhance the capacity of universities for contributing to climate action, or exacerbate the negative impact, or both simultaneously?

Internationalisation of higher education manifests itself in many different ways, including attracting students from elsewhere in the world, international mobility of home students, an international staff body, diversified curriculum, research collaboration, global research focus and the global impact of research, among many others. There has been extensive attention paid to categorising the diverse forms of internationalisation, with the most prominent distinction made between 'internationalisation at home' (incorporation of global perspectives into the curriculum) and 'internationalisation abroad' (student and staff mobility, and transnational education) (De Wit and Altbach 2021). While making use of this distinction, for the purposes of this chapter another categorisation will be introduced: that between *actors* – who is involved?; *practice* – what is happening?; and *influence* – who or what is affected?

Internationalisation involves diversification of *actors*, the people involved in universities – most obviously the student body but also the make-up of academic staff, the collaborators in a research project and the bodies commissioning consultancy. Internationalisation also involves changes to the activities carried out and the way in which they are conducted (*practice*). In the curriculum it involves a diversification of subject matter and epistemic framing, while in research and community engagement it involves changes in the geographical focus of activities. Internationalisation also involves changes in the *influence* of those activities. The graduates of an institution may return to their home countries and take with them the attributes acquired, applying them in their professional, civic and personal lives. Innovation and basic research are also disseminated at a global level to an increasing extent.

While these three elements are intimately linked in practice, it is analytically useful to separate them out, as they exist in varying configurations and with different implications (both generally and specifically in relation to climate change). For example, a university can have a largely national student body, but still incorporate an internationalised curriculum; conversely, a university may have a highly internationalised student body for revenue generation purposes, but maintain a narrow nationally focused teaching and learning experience. In relation to research, some institutions achieve global reach with their findings and publications (on account of the high standing in which their researchers are held) despite not having close engagement with issues and researchers in other contexts.

These questions relating to actors, practice and influence are ones that we can pose of any internationalisation initiative. The choices for each of these will have an impact on climate change, either through direct impacts (emission of greenhouse gases, mitigation or regeneration) or indirect impacts (through influences on individuals, society and technology). It is important to emphasise from the outset that internationalisation is not an either/or. No universities are exclusively local or national, and none are purely international. There is a continuum of internationalisation, or perhaps continua, operating in parallel in relation to the different dimensions of student body, curriculum and so on). This section will assess the influence that the movements along the continua of internationalisation have on climate action through each of the five modalities of education, knowledge production, services, public debate and campus operations.

## Education

The education modality shows the contradictions involved with internationalisation well. On the one hand, it is a prominent cause of greenhouse gas emissions, and on the other it represents the most powerful potential means of transformation towards a sustainable world. The most obvious implication of internationalisation of higher education is student travel.[2] The greenhouse gas emissions arising from transport (predominantly air travel) are the most tangible and best documented of all impacts of internationalisation. This area is also particularly hard to address, as it is bound up with a vital source of income for many universities – as much as 45 per cent of income in research-intensive Russell Group universities in the UK, for example (Shields and Lu 2023). There is also mobility of academic staff for the purposes of teaching, either between different institutions or among the international branch campuses (IBCs) of a single organisation. Internationalisation of actors is, therefore, predominantly negative in terms of its direct impacts. There have, however, been efforts in a number of institutions to mitigate these impacts, through moves to online education (reducing student travel) and online conferences and meetings (reducing staff travel), as well as carbon offsetting and broader moves towards net zero campuses. The establishment of the Climate Action Network for International Educators (CANIE[3]) and its accord for HEIs (CANIE 2022) shows the increasing awareness of the need for concerted action in the sector.

In addition to these direct impacts, there are ones occurring indirectly – through the value of a diverse student body and the linkages with internationalisation of the curriculum and global influence. Internationalisation of the student body brings a dramatic change in the influence of universities through their graduates, as – even allowing for a proportion who remain to work in the countries in which they studied – most will return to their countries of origin or to a third country, thereby contributing to the global circulation of ideas and skills. Having diverse perspectives in the classroom also aids understanding of the complex nature of climate change, involving difficult compromises between the interests of different countries, while for those travelling it can be a life-changing experience fostering more pro-environmental attitudes (Campbell et al. 2023; Wynveen et al. 2012). These outcomes are not inevitable but depend on how internationalisation is curated: in some cases, the international student body may be fairly homogeneous or, if diverse, that diversity may have little impact on academic culture, with other knowledge traditions subordinated to mainstream, Western ones.

In terms of practice, many publications have advocated for internationalisation of the curriculum, intercultural understanding and global citizenship (for example, De Wit and Altbach 2021; Ilieva and Tsiligiris 2021). In relation to climate change specifically, this diversity of perspective is crucial given the inevitably global nature of the issue, and the need to find new forms of thinking to address complexity and to draw on the resources of Indigenous knowledge (McCowan 2021b). One of the rationales for internationalisation (for example, Qiang 2003) is to equip graduates for the global workplace, interacting with people from diverse cultures and languages. This rationale is also relevant to climate change, given the need for international working. However, some of these forms of internationalisation are pernicious in relation to climate change – if they are wedded to the paradigm of global employability at all costs outlined in the previous section, prioritising individual career advancement and corporate profit above public good and environmental protection. In some cases, these aims are embedded within marketing strategies and explicit graduate outcomes (sometimes problematically merged with notions of global citizenship), but in others they are implicit and unwittingly supporting a predatory transnational elite.

## Knowledge production

Internationalisation of research also has divergent implications. On a positive note, it is widely recognised that international collaborations in research are highly esteemed and can also lead to impactful and high-quality outputs (Gazni et al. 2012). As emphasised above, in relation to climate change these forms of collaboration are particularly important, given the global and multifaceted nature of the phenomenon. These positive benefits stem not only from the *actors* component, but also the *practice* and *influence* ones, in engaging with diverse contexts in research and innovation, including comparative perspectives. Although rarely achieved in practice, these forms of collaboration can also display an 'ecology of knowledges' (Santos 2015), when juxtaposing and bringing dialogue between diverse epistemic perspectives, languages and worldviews. International dissemination of research naturally increases its reach and influence, and also enables a global dialogue.

However, there are also some pitfalls in the internationalisation of knowledge production. While not reaching the levels of student mobility, the transport of academic staff through mobility, fieldwork and conferences is also a significant cause of direct emissions (Bjørkdahl et al. 2022; Gill 2021). These forms of travel are not without their value in research terms, so there will inevitably be some trade-offs. There are also potential indirect downsides: international research collaborations have historically been dominated by elite institutions located in the Global North, at their worst reducing partners in the Global South to simply data collectors, while analysis of collaborative publications shows that they too are dominated by these elite institutions (Gazni et al. 2012). While more attention is now being paid to genuine partnership and avoiding asymmetrical relationships, these patterns are still entrenched (Grieve and Mitchell 2020; Oldac et al. 2023). With the global science and publication systems heavily weighted in favour of high-income English-speaking countries, there is a danger of non-central cultures being reduced to the 'local' and marginalised from the debate. These potential drawbacks are not inevitable, however, and should not dissuade universities from engaging in diverse forms of international collaboration in knowledge production.

But just as we need to move from mitigation and adaptation to regeneration in environmental terms, so we need to move from simply avoiding or reducing negative forms of research collaboration to creating new transformative ones. As will be explored further in Chapter 9, co-production with diverse communities and participatory action research can both empower those usually marginalised from knowledge production processes and enhance the quality and relevance of research, putting in place the conditions for social and environmental regeneration (Frediani and Nussey 2021).

## Services

The services modality can involve a range of different activities, and varies dramatically from institution to institution. In the case of large institutions corresponding to Clark Kerr's (1963) 'multiversity', these will be extensive and involve hospitals, spin-off firms, sporting and artistic facilities for use by local communities, environmental services, wide-ranging consultancy and secondments of staff. Not all of these will be significantly influenced by internationalisation and may continue to take place within a largely local and national remit, particularly if they require physical access (for example, a health clinic). But internationalisation leads to some services being provided globally, such as consultancy and technical support conducted at the international level, facilitated by developments in information and communications technology (as well as increases in international travel).

These varying forms of technical support might have negative consequences in involving unnecessary flying of experts around the world, with direct and indirect influences: direct in terms of the emissions, but also indirect in undermining local expertise and reinforcing dependency. The positives include cross-fertilisation of ideas in the application of knowledge between diverse contexts, and also between diverse sectors of society.

## Public debate

Extending the public debate role of universities internationally is vital in relation to climate change, with awareness-raising, deliberation over possible solutions, considerations of justice and learning from alternatives all needing to take place across national boundaries. Developments in technology, in particular the internet and social media, have meant that there are real opportunities for making public engagement global, although some fora will still remain local and national. Cross-cultural communication and debate is facilitated by linguae francae, in particular English, although herein lies a risk of asymmetrical opportunities for participation, privileging those who either have English as a native tongue or have had high-quality education in the language, and marginalising others. There are a number of other factors that also hinder the equal participation of certain groups in debate, even in conditions of formal equality, so attention is always needed to address barriers to voice, expression and contribution – both as a question of justice and in order to maximise the possibilities of creative solutions being found.

## Campus operations

In most cases, campus operations are by definition a local matter. While internationalisation may bring international students and staff onto campus – increasing the total number of people and therefore the environmental impact – the traditional campus remains bounded in physical, local space. Greenhouse gas emissions, of course, are inevitably global, so this influence is and always has been internationalised, regardless of the actions of the university.

However, there are some new developments that affect how we understand campus operations. The development of online education has reduced the emphasis on the physical campus, and in some cases has led to entirely virtual institutions. The environmental implications of these virtual campuses are distinct: while there is a significant reduction in transport costs, the energy usage of powering the computers involved must be factored in. Another development is the increase in transnational education of a face-to-face nature, primarily through IBCs. Management of a network of international campuses may significantly decrease international travel, but may bring some environmental costs, particularly if it involves mobility of staff and students from the 'home' campus.

While we think of campus operations primarily in terms of negative impacts – across Scope 1, 2 and 3 emissions – there is the possibility of transformative action, in the form of regeneration. While this process may occur independently, there are ways in which internationalisation can enhance the human and environmental forms of regeneration, in a similar way to ecotourism – as argued by Lee and Lundemo (2021). Furthermore, internationalisation can enhance campus sustainability by bringing an awareness of the global ramifications of local actions.

*

The sections above have outlined some of the prominent implications of the internationalisation of higher education for climate change. Not all of these will apply to all institutions, and there will inevitably be others not listed here; institutions should carry out this mapping process in accordance with their specific contexts. It is also important to highlight that – with the exception of outcomes in the form of measurable greenhouse gas emissions – categorisation into positive and negative is challenging, with many activities having shades of both and needing nuanced treatment. Furthermore, the complexity of the climate system (and the higher education system) is such that many impacts are unpredictable, especially in the medium to long term, so flexibility and constant monitoring and revision are required.

An important task, therefore, is for universities to carry out their own diagnosis of activities. Figure 5.1 represents a kind of diagnostic tool that might be used, using the example of the education modality. For this figure, the direct impacts are listed on the left-hand side, and the indirect ones on the right, with the positive impacts in the top half, and the negative ones below. The impacts are coded in relation to the dimensions of actors, practices and influences. As stated above, there is no neat binary between positive and negative impacts, and the reality is one of a continuum; many activities will have aspects of both but will veer one way or the other.

The absence of direct positive influences in the case of education might be disheartening for many working in the field. However, this is just a result of the nature of the educational process: educating people in climate does not in itself reduce emissions or regenerate ecosystems,

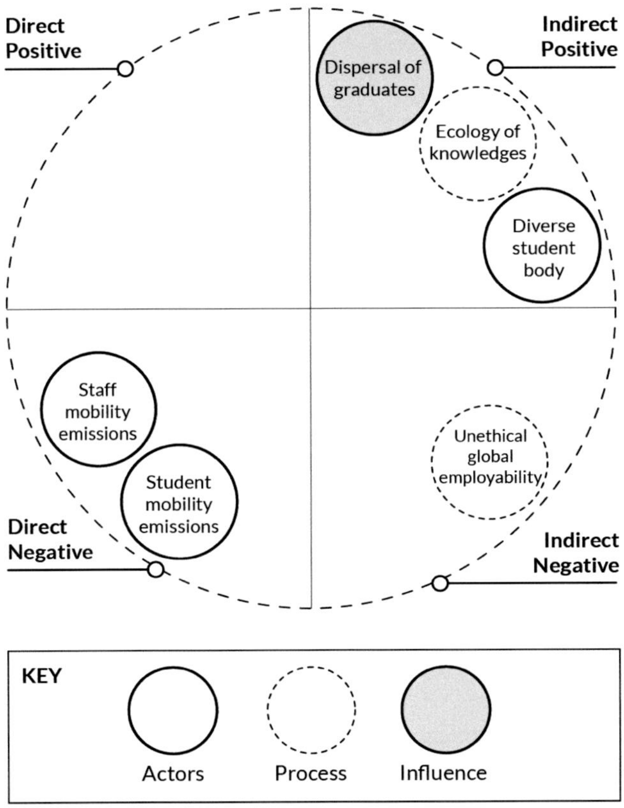


**Figure 5.1**: Internationalising the education modality.
Source: The author.

it does so indirectly. Fortunately, the indirect impacts are multiple and extremely significant, so education remains a fundamental part of any climate action strategy.

Inevitably, looking at this diagram (and indeed looking at the central contradiction of internationalisation of higher education), there will be an urge for comparison between different activities. A university might pose the question: if we maintain our high numbers of international students and increase our research into clean energy usage in Africa, will the latter cancel out the former? In most cases, these kinds of comparisons cannot be made precisely. The direct emissions of various activities can be compared, but rarely the indirect impacts. In some cases, the issue is the difficulty of quantitative measurement; in others, the impact is diffuse or occurring over a long timeframe, making difficult any kind of assessment (McCowan 2022). Cordero et al. (2020) attempt with some success to convert university learning to tonnes of carbon dioxide emissions, although not all indirect impacts will be so amenable. It is possible to trace the graduates of a module but not all members of the audience of a radio programme, for example.

Further challenges of comparison are raised when we are dealing with regeneration (Lee and Lundemo 2021) – the actions needed in addition to decreasing greenhouse gas emissions (mitigation) and finding ways of living with changes in the climate (adaptation), to transform our communities and natural environments for a flourishing future. Nevertheless, while we must accept that some impacts occur but can never be adequately gauged and attributed, there are creative ways in which comparisons can be made, trade-offs understood and holistic assessments produced. In any event, while some prioritisation will always be necessary, universities will normally need to be working in all of these areas simultaneously.

## Navigating global dynamics

The discussions in the previous section have to a large extent assumed impact to be a unitary force emanating from a university (although with potentially positive or negative effects on society and the environment). Yet impact can vary in significant ways. Three of these ways are particularly significant: the two dimensions of intensity and timescale (outlined in the previous chapter) and, a third, reach. First is the *intensity* or strength of the influence – for example, whether the impact of learning about renewable energy alternatives has been relatively superficial, or profoundly life-altering and leading to a permanent shift away from fossil fuels. The second is the *timescale* or speed with which the impact takes place. There is significant variation in this dimension: some influences are immediate – for example, the publication of a groundbreaking discovery about ice loss, accompanied by extensive news coverage and policy debate; and others are very gradual – for example, the uptake of electric cars across a population. The final dimension is *reach*. This dimension indicates the extension of the influence, in other words the number of people affected and the spread across locations and countries. The element of scale is acknowledged to some degree in the movement between the stages of bridging actors, society and ecosphere, in the movement from the more particular and local to the more diffuse. Yet there is still substantial scope for variation within each of the stages.

These three dimensions interact. Some impacts will provide a subtle influence over a large proportion of the population, while others will lead to a major transformation but in a restricted number of people. Normally, one would expect an inverse relationship between intensity and reach – so the more focused an impact is, the stronger it will be, and the more dispersed the weaker (although this is not always the case). Some university impacts will be low intensity but act over a long period of time, whereas others will be strong and short-lived. In some cases, the level of intensity will change over time, either growing stronger or weaker. There may also be interactions between reach and timescale, with influence becoming broader over time: although there are certainly exceptions, particularly in the internet age, in cases of almost instantaneous global uptake of an idea.

What affect, then, does internationalisation have on these dynamics of impact? The previous section has analysed the ways in which internationalisation manifests itself within the five modalities, and knock-on impacts on climate change. Intensity, timescale and reach are closely bound up with the modality in question: so, for example, the timescale for the impact of campus operations is usually more immediate than that for research, and the reach of a legal clinic is usually more restricted than a social media post. Internationalisation, therefore, operates within the bounds of these modality-specific dynamics. However, there is an obvious influence on the third dimension *reach* through internationalisation, as it brings an expansion in actors, practices and influence. In this way, universities extend the reach of their teaching through IBCs and distance programmes, while the influence of their international graduates extends potentially throughout the globe, and

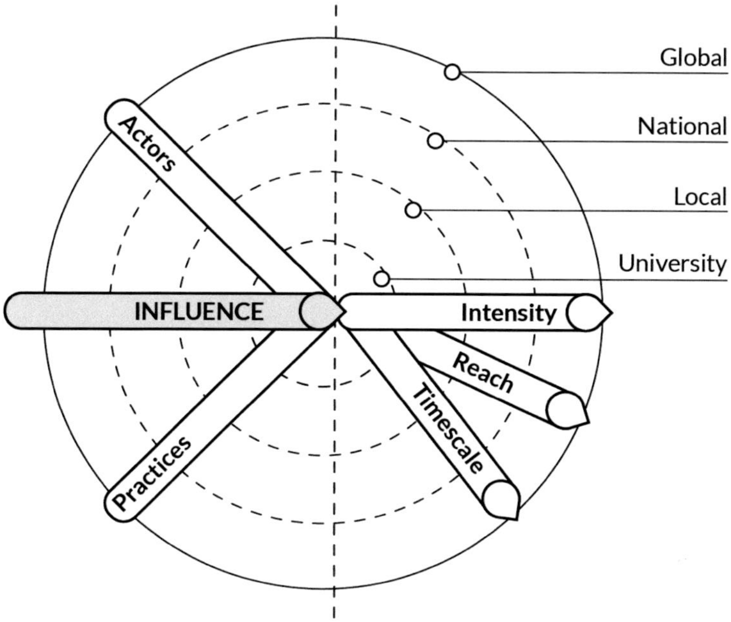


**Figure 5.2**: Dynamics of impact through internationalisation.
Source: The author.

the partnerships forged with institutions in other continents and the dissemination through international journals bring their research to global audiences.

These dynamics can be seen in Figure 5.2. Internationalisation is constituted by changes in actors and practices, and then via the influence of the five modalities, moving between local, national and global levels. Influence demonstrates varying levels of intensity, reach and timescale.

Universities vary in the focus of their engagement, with some having intense impact in their local area with less visible global impact, while others focus on their global reach but may have less contact with their local communities. All universities have concentric circles of influence, although they may be more strongly concentrated close to themselves, or more distantly. Does global reach spread out gradually from university to local, national and global, like ripples from a pebble thrown into a pond? Some forms of influence may well move like that, in the vein of Malcolm Gladwell's (2000) 'tipping point', describing the movement of viruses, ideas, behaviour and fashion. Yet it is possible for global reach to bypass the local and even the national in the internet age. This is the case with social media engagement, where an academic's tweets may have repercussions on the other side of the globe but not in the local community, or where a university may be publishing and disseminating its research for global not local audiences – both of these dynamics intensified when the language (for example, English, Spanish, French) is not spoken by the local population.[4]

This figure can again serve as a mapping tool for institutions and systems of higher education in understanding the influences of their activities across different scales. In contrast to reductive notions of impact, it is important to acknowledge the potentially long-term ramifications of learning and scholarship, the varying intensity and reach of influences, and the multidirectional movements between university, local, national and global.

These considerations also lead us to reflect more broadly on the role that universities play as part of a global system (independent of their deliberate internationalising actions). Even though the global higher education system is neither an integrated nor a coordinated one, the distribution of institutions, their distinct functions and the way in which they interact with each other all have a strong impact on climate action.

There are interactions of influence that take place without any deliberate intention, and even without any knowledge of those taking part. These actions might be reinforcing: for example, Chinese and South African universities may both be working with local communities in Mozambique to regenerate coastal defences and ecosystems, and the ideas and practices may combine to create a stronger (intensity) or broader (reach) or more extended (timescale) influence. This combination may be simply one of aggregation – of 'adding up' positive influence – or the interventions may actually be interacting and combining in dialogue to create new configurations. In other cases, they may be in tension or contradiction – say, for example, if China offers a more technical approach, bringing in external experts and offering solutions based on technological innovation, and South Africa offers more participatory contextualised approaches drawing on the knowledge of local actors. We can add to these many other examples from the education modality (the diverse impacts of graduates from different institutions and different courses) and the knowledge production modality (the interaction of ideas and innovations). These interactions are extremely difficult to chart in practice but must be acknowledged.

However, there are also intentional ways in which universities combine. Academics very commonly collaborate across institutions, and across national boundaries, for the purposes of research and publication.

Sometimes, these interactions solidify into formal partnerships and programmes of work between institutions. Although challenging in practice, these partnerships allow institutions to capitalise on the potential synergies outlined above and combine different areas of expertise and experience to advance understanding and provide more effective interventions. Sometimes, these partnerships grow into consortia involving multiple institutions, even in different continents, working on large-scale programmes of work. And finally, at the largest scale there are international alliances and associations of universities. These rarely entail specific programmes of work but provide the space for the alignment of goals, pooling of expertise, inter-institutional learning and reinforcement of commitments. These alliances are common in the area of sustainability and climate change, with examples including the Global Alliance of Universities on Climate and the Association of University Leaders for a Sustainable Future.

It goes without saying that these international interactions, while hopeful and generative, act within the conditions of possibility outlined in the first section of this chapter. Disparities of resources and prestige, as well as constraints on the orientation of universities' activities to the public good, limit the possibilities for generative global interaction in addressing climate change. There are, needless to say, complex normative considerations about what universities *should* do, given this global backdrop – ones that will be explored in the second half of this book. These choices are even more crucial for highly visible institutions in the Global North, as their actions serve not only to bring major benefits for humanity but also to constrain the actions of less visible and less well-resourced institutions through supporting the pernicious mixture of homogenised form and unequal opportunity, rather than diversity in a horizontal system.

## Notes

1 Including imputed student costs.

2 It might also be possible to categorise student mobility as part of the campus operations rather than the education modality, since it is not a result of the teaching and learning process itself, but relates to the location of the campus.

3 https://canie.org/

4 Nevertheless, there can be waves back: so the research published in English may end up having an impact on the local community via global and national policies, practices or technology development.

# 6
# Teaching climate change

It is hard to argue against addressing climate change in some form in higher education. There is a pressing need for all people, in the professional, civic and personal dimensions of their lives, to be informed about the causes and effects of climate change, to have the critical capacities and imagination to forge alternatives for the future, and to have the skills and commitments to bring change individually and collectively. Professional and general education in today's world cannot ignore the issue, given its increasing centrality in our lives, with the changes under way in the Earth's atmosphere, on its land and in its oceans, and the impacts these have on flora and fauna, human health and well-being, and the political, economic and cultural dimensions of human societies.

Yet a clear mandate for the inclusion of climate change in higher education does not come with a clear recipe for how that should happen. Should we focus on ensuring that all students acquire the body of scientific knowledge on the changing climate and its causes? Should we equip all students with a set of climate-related competencies or skills on finishing their degrees? Should we provide a space for reflection on the moral and political dilemmas raised by the climate emergency? And what approaches to teaching and learning can best support these different aims?

Climate change in fact presents some particular challenges for lecturers in higher education. It is relevant to all disciplinary areas, although many non-specialists may feel themselves ill equipped to incorporate the material into their classes. It has highly theoretical dimensions and large bodies of scientific evidence, but it is also a practical issue, involving changes to lifestyles and governance. It is also highly contested, and likely to fuel disagreements and even conflict in the classroom. These challenges may well discourage many of those teaching in universities from incorporating the issue in their teaching.

This chapter, however, will adopt a different standpoint. While not disregarding the above concerns, it argues that climate change in fact represents a driver for positive change in higher education, in spurring us on to bring shifts in teaching and learning that are long overdue and to enhance the transformative potential of the experience.

There is a growing body of literature on climate change education, drawing on the longer traditions of environmental education and education for sustainable development (for example, Anderson, A. 2012; Bangay and Blum 2010; Bryan 2020; Facer et al. 2020; Kwauk and Casey 2021; Oberman and Sainz 2021; Tannock 2021). While most of this literature relates to the school level, there is an encouraging increase in accounts relating specifically to higher education (for example, Bush et al. 2017; Bussey 2010; Facer 2020; Fahey 2012; Nussey et al. 2023; Senbel et al. 2014). Nevertheless, the literature is still dominated by specifications of what students should know and be able to do (whether conceptualised as climate knowledge, green skills or sustainability competencies), with an assumption that it is a straightforward step for educators to foster those capacities in students. It is important to address the pedagogical question head on, by valuing and aiming to understand the *process* as well as the *outcome* dimensions. Reflecting on processes – the moment of teaching and learning itself, and the interactions between students and lecturers – is important partly to provide those teaching in universities with the tools to shape their own practice, but also to leave the door open for students to construct their own pathways and outcomes.

This chapter should be read in conjunction with the following one, Chapter 7, focusing on the location of climate change within the higher education curriculum. Chapter 7 argues that universities need to approach their provision for climate change as a *topography*, involving not only formal, taught courses (classroom), but also student-led initiatives in the university (campus) and experiential learning beyond it through work experience and mobilisation (community). This chapter focuses specifically on the first of these three spaces – the teaching of climate change in the 'classroom' – understood not only as the literal room but any space for formal (and most likely accredited), taught provision. It explores the possible ways to engage with climate change in the classroom – given its complex epistemic nature – and the implications that they have for student learning and society's ability to address the crisis. Inevitably, the questions of 'what' to teach and 'how' to teach are intertwined, and there are many crossovers between the curricular and pedagogical discussions.

While focusing on pedagogy, and on process as well as outcomes, this chapter does not purport to be a comprehensive 'how to' on teaching climate change. There are many relevant approaches and techniques for engaging learners, fostering critical questioning and for the collaboration and application of knowledge that are covered in the broader literature on teaching and learning in higher education (for example, Ashwin et al. 2020; Fry et al. 2008), as well as developments in digital technologies (for example, Laurillard 2002) that cannot be dealt with in depth here. Furthermore, there will need to be contextualisation within particular locations and disciplinary areas. The focus here is on those specific aspects of climate change that can – possibly unexpectedly – be a spur to positive pedagogical transformation.

## The response of education to the climate crisis

Not all responses to climate change require education. Some commentators (for example, Gates 2021) hold out faith in technological developments, either in global responses on a massive scale such as geo-engineering (including atmospheric sprays or subterranean carbon capture), or more dispersed innovations such as electric cars and new forms of energy generation. These responses would require only specialist professional and scientific education for the few. Most, however, recognise that simply waiting for these developments is hazardous to say the least, given the urgency of the threat, and assert that broader awareness and practical action is needed across all populations. Furthermore, the development of new technology, and the orientation and usage of existing technology, are strongly bound up with economic, political and cultural currents, and therefore we cannot easily separate the two.

The various conventions and agreements relating to climate change and sustainable development invoke education as part of the means of rectifying the situation. Article 6 of the UNFCCC (1992) requires states to promote and facilitate 'the development and implementation of educational and public awareness programmes on climate change and its effects', 'public access to information on climate change and its effects' and 'training of scientific, technical and managerial personnel'. Education, training and public awareness have been built into subsequent climate agreements, including Goal 13 of the SDGs (Reid 2019). There has in fact been significant change at the official level, with only 5 per cent of UNFCCC signatory countries not incorporating climate change into their curricula at any level (UNESCO 2019). Mexico and Italy have made climate change education compulsory throughout all levels of the education system (Kwauk and Casey 2021). Yet very often this official adoption has not translated into adequate learning, on account of insufficient teacher preparation and the marginal position of climate change education in the curriculum (Reid 2019).

The logic underpinning the role of education in addressing climate change is conventionally that awareness leads to action: people do not know about climate change, or do not understand it sufficiently, and that is why they are not doing anything about it. This logic has two important characteristics: first, it is a deficit model in assuming lack of knowledge and understanding; and second, it takes as read that awareness will lead to action. Both of these are problematic, as much in their factual basis as in their educational implications.

Research has, in fact, shown the dubious efficacy of awareness-raising of a purely cognitive nature (Anderson, A. 2012; Bangay and Blum 2010; Facer 2020; Facer et al. 2020; Monroe et al. 2019; Rousell and Cutter-Mackenzie-Knowles 2020; Stevenson et al. 2017). Yet moving beyond the knowledge element is demanding for educators. Teaching climate science requires little departure from traditional formal education, but challenges arise when moving into the realm of responses to the climate emergency:

> The distinction between 'just the facts' and 'also the actions' may separate some science educators from environmental educators, but also may highlight the point at which educators believe a fundamental science topic becomes political, and therefore too close to advocacy for classroom educators to address. (Monroe et al. 2019: 792)

Aiming to address the gap between knowledge and action, many educational and public information approaches focus on behaviour change. Largely targeting the individual level, they point learners towards practical shifts in their patterns of consumption and usage, and contributions to regeneration at the local level. While these are all laudable actions, they are only part of the change that is needed in order to address the climate emergency, which also requires attention to corporations, governments and underpinning societal structures. In focusing only on the individual, these educational approaches leave the political and economic roots of climate change unexamined. We can therefore see that a series of balancing acts are needed by educators, in moving from knowledge to action, from the individual to the collective and from the technical to the political. This shift to the other pole in each of these three binaries does not exclude the original pole, however, which still retains its importance (in other words, knowledge of climate change will always be essential, even though not sufficient).

As part of this balancing act, a number of alternative approaches have emerged in recent years. These have included engaging students more effectively through incorporating experiential elements, either with simulated experiences or real-life environmental challenges, and the use of digital technologies and social media (Senbel et al. 2014). Other approaches have focused on the emotional or affective (Ojala 2016; Bryan 2020; Lehtonen et al. 2019), and argue for psychosocial or eco-social approaches, learning through artistic or cultural activities or through activism and campaigning (Nussey 2021). The idea here is that emotions can be turned from being a stifling force to a liberating one. The intertwining of the human and the natural environment realms leads to an approximation of climate change education with citizenship education, as argued in Jacobi (2003).

Kwauk and Casey (2021: 46–7), for example, aim to connect with students in this more holistic way through their frame of five 'design elements' of climate change education:

1. A cognitive point of entry, like a recent climate-related disaster or the introduction of a local environmental resource challenge, and the possible solutions to it;
2. An affective dimension that helps to cultivate one's empathy toward the environment;
3. An existential component that challenges one's sense of self, one's way of living and being, and one's values, beliefs, and worldview;
4. An ownership dimension, like the self-identification of a community-based environmental challenge, or developing a plan that builds one's knowledge of, personal connection and commitment to, and sense of responsibility for a local climate change issue;
5. Opportunities for empowered action or dissent, like a community action project or political protest to not only feel a sense of direction, but to also increase sense of agency and confidence in making informed choices.

Cordero et al. (2020), similarly, identify the three key attributes of empathy, ownership and empowerment to promote through their curricular programme, encompassing the elements of climate science, climate solutions and environmental communication.

Despite these drives towards more experiential, affective and empowering learning, much climate change education continues along narrower cognitive lines. These approaches not only display different pedagogical methods but cluster around fundamental differences of political, moral and epistemic values. In their discussion of the 'new green learning agenda', Kwauk and Casey (2021) distinguish between three approaches to climate change education: (i) skills for green jobs (employment focus); (ii) green life skills (personal and civic as well as professional, lifelong and life-wide); and (iii) skills for green transformation (addressing oppressive structures and political transformation). This frame shows the continuum from more technical to more political, and narrower to broader conceptualisations of the scope and impact of climate change education. As seen above, government approaches tend to cluster around the transmission of information and more technical training (Nussey 2021). At the broader and more political end are the transformative approaches drawing on the ideas of Paulo Freire, which aim for collective political action through *conscientisation*, as well as approaches based on feminist and decolonial thought, the development of a critical understanding of injustices and action to forge a new form of society (Facer 2020; Lotz-Sisitka et al. 2015; Macintyre et al. 2018; Sterling 2011; Tannock 2021).

As stated above, climate change education builds on the older tradition of environmental education and the more recent education for sustainable development. While anthropogenic global warming on the basis of greenhouse gases involves a particular set of facts and body of scientific knowledge, many of the underpinning issues are similar to earlier environmental challenges – involving complex moral questions of how human beings can live with each other and within the carrying capacity of the ecosystem. This long tradition of environmental education provides a wealth of experience, inspiration and methodological tools for contemporary climate change educators, although many of the challenges and tensions are still there. All instances of what is sometimes called 'adjectival' education (cross-cutting curricular themes, such as citizenship, diversity, human rights or peace) suffer from the lack of a defined disciplinary space, dedicated professionals and time allocation. Incorporating these cross-cutting themes in higher education can be particularly challenging, as explored in the section that follows.

## Trends in higher education as a learning space

As seen above, global agreements on climate change have advocated for education at all levels of the formal system, in professional training and in public awareness campaigns. Few would disagree that all stages and forms of education need to be harnessed in such a critical challenge. While there are some commonalities, each of these levels involves different teaching and learning approaches. Higher education presents some particularities on account of the following features: in contrast to school, it is voluntary and non-universal in most countries; it is highly specialised in nature, usually in a single disciplinary or professional area; teachers in higher education are usually researchers and scholars in a disciplinary area, and rarely have extensive pedagogical preparation; and there exist significant areas of institutional activity alongside teaching – most importantly research, but also community engagement and consultancy. In addition, while the composition of student bodies differs dramatically from institution to institution, in many cases universities show diversity in terms of the origin of students, providing a space for interaction across regional, national, cultural and linguistic groups.

Three areas of change in higher education are particularly relevant to the question of teaching of climate change, and that of location within the curriculum, addressed in Chapter 7. The first is the oft-mentioned phenomenon of massification, equating to rapid increases in enrolment across the globe – although still with significant differences between countries (low-income countries average 9 per cent gross enrolment ratio, lower-middle-income countries 27 per cent, upper-middle-income countries 38 per cent and high-income countries 79 per cent; UIS 2022). This process of expansion has brought new social groups into the university, in many cases ones that have not had high-quality schooling and face particular challenges with the academic demands. Massification has also led to growth in the number and diversification of type of courses on offer at universities, with an increase in vocational programmes. With an expanding proportion of the population in the university, the civic role of higher education has become more prominent, and this level of education has increasingly been used for promoting social goods – climate change education being a clear example – although there are many who question whether this is an appropriate role of the university (Martin 2022).

In parallel with the process of massification, universities in the twentieth century and beyond have endured a period of epistemic tension and conflict, as discussed in Chapter 3, one that has thrown into question the institution's worth, authority and legitimacy. Philosophical currents of poststructuralism have questioned the moral and epistemic certainties of the Enlightenment and unsettled the position of the academic community as the bastion of truth and knowledge. These influences, combined with the encroachment of neoliberalism – which has co-opted the postmodern decentring for its own ends – have led many commentators to point to the 'university in ruins' (Readings 1996), or the crisis of the university (Santos 2004). Technology has also exacerbated this perceived crisis in various ways: in undermining the university's role as a literal store and archive of knowledge by making information broadly available through the internet; in challenging the role of the professor through the processes of unbundling and standardisation of content (McCowan 2017); through the emergence of massive open online courses (MOOCs) breaking universities' monopoly on taught provision; and with the increasing prevalence of learning analytics and surveillance mechanisms (Laurillard and Kennedy 2017; Williamson 2018). These currents have had a political dimension, in being mobilised in the quest for recognition of marginalised groups and in challenges to patriarchy, social class elitism, white supremacy and heteronormativity. Universities have been strongly bound up in the 'culture wars', and the unsettling and reinstatement of the Canon (Bloom 2014), and battles between defenders of the Western Enlightenment (for example, Pinker 2018; Peterson 2018) and the 'woke' generation. As mentioned above, the populist movements of 2016 represented a further challenge to universities' epistemic legitimacy (Ignatieff 2018). The Rhodes Must Fall movement in South Africa set in motion calls for decolonisation that have spread around the world, even to the former colonial centres, and have brought new awareness of the 'monoculture of the mind' in universities and the need for epistemic pluralism (Del Monte and Posholi 2021; Mutanga and Marovah 2024; Santos 2015; Shahjahan 2015; Shiva 1993).

Another quite different influence on universities relevant to this debate is the increasing attention given to the quality of teaching and learning. With some exceptions – both institutionally and individually – universities have historically neglected teaching quality in favour of research quality, and in contrast to schools have not required their staff to have any form of teaching qualification. In recent decades, marked changes in this area have been evident, spurred on by the processes of massification outlined above, and the arrival of non-traditional students with different learning needs, the increase in the number of teaching-only institutions, the need for universities to demonstrate their teaching and learning quality in the context of marketisation and consumer choice, and value for money demands on universities to account for public funding. Universities around the world have set up teaching and learning units, introduced mandatory courses for staff – sometimes accredited – and provided other forms of learning support for students.

Research and scholarship on teaching and learning in higher education also grew markedly in this period, in some cases showing the disappointing learning gains of students in traditional higher education (Arum and Roksa 2011; Schendel 2015), as well as pointing to the efficacy of particular methods (Kember and Gow 1994). While there is contestation within the field of scholarship of teaching and learning, as in any field, some common emphases have emerged: reduction in the use of the traditional lecture; use of the 'flipped classroom'[1] as a way of ensuring content delivery does not crowd out the space for discussion in precious class time; introduction of collaborative group work; movement in lecturers from a knowledge transmission to a learning facilitation role; replacement of final exams with coursework and portfolios; and introduction of problem-based and case-based learning (Ashwin et al. 2020; Biggs 1999; Kember and Gow 1994; McCowan et al. 2022). Developments in digital technologies have also brought significant changes to the teaching and learning experience in universities – with an acceleration in this process due to the Covid pandemic.

Furthermore, universities have paid increasing attention to their broader educational responsibilities, beyond specific disciplinary areas. This trend has been underpinned by drives for employability, in part through market mechanisms of choice, but also the pressures of government in their designs for a highly skilled workforce. Commonly now, HEIs advertise their list of graduate attributes, variously including qualities such as global citizenship, intercultural competence, leadership, teamwork, critical thinking, problem-solving and digital literacy – many of which are relevant to climate change. These shifts, however, are often a target of resistance by academics, associated as they are either with a challenge to traditional disciplinary practices or with a Trojan-horse-like neoliberalisation of the curriculum.

While the above trends have made possible the incorporation of climate change into the higher education curriculum, at first sight its inclusion seems to involve some significant problems. First, there are issues internal to the university that militate against its effective inclusion. Climate change is not located within a single discipline and encounters significant obstacles to either embedding itself across all courses or establishing itself as a discrete unit – in light of the autonomy of lecturers to design their courses, curriculum overload and professional accreditation. There can also be resistance to value-based initiatives and cross-curricular provision from both staff and students.

Second, teaching climate change is complex because of the nature of the issue. It is a contested question, with a range of possible positions cohering with scientific evidence, not to mention vocal challenging of the science from some quarters. As a controversial issue involving deep moral and political values about the way our lives and societies are organised, it is challenging to navigate in the context of a diverse classroom. Arts, humanities and social sciences should be more comfortable with these kinds of moral debate and value complexity than natural sciences and STEM subjects with a more numerical basis, but they still face distinct challenges in integrating issues of climate change, which may appear to be the job of environmental science courses. Climate change also involves a range of skills and practical experiences that are difficult to develop in the classroom context.

Third, climate change is embedded in the set of deeper epistemic challenges in contemporary societies that are outlined in Chapter 3. Universities have long been centres of struggles for meaning and legitimacy, in the context of postmodern challenges to Enlightenment notions of objective truth and progress. The traditional theocratic views that the Enlightenment had displaced in Europe have also had a resurgence in different parts of the world, providing an unexpected ally with those challenging the Enlightenment project. Populist leaders have taken advantage of the stand-off to turn the epistemic vacuum to their advantage, making use of social media to spread fear, prejudice and disinformation. Climate change has been positioned by these populist movements as a smokescreen for socialism, big government and threats to individual liberties, bolstered by the deliberate campaigns of misinformation from fossil fuel lobbies.

Nevertheless, despite this unpromising backdrop, this chapter will attempt to show that climate change is in fact a spur for much-needed change in higher education. While highly challenging, if we do succeed in addressing it in university teaching and learning, then we will have achieved a transformation that was needed in any event. This dynamic has been expressed in relation to the broader societal shifts required by the climate crisis. So, Helm (2020) refers to 'no regrets' policies, ones which help reduce carbon emissions or increase carbon absorption, but will achieve other desirable social and environmental ends at the same time. Naomi Klein's (2014; 2019) work in this way emphasises the inseparability of the social movements for workers' and Indigenous rights, and the environmental movement, given climate injustice and the inseparability of social and natural exploitation.

In the field of education, the inclusion of climate change can signify an opportunity to engage more deeply with the fundamental questions of human existence that cut across all disciplinary areas and university courses, and to enable the transformation of students in higher education across personal, civic and professional domains – in short, the aims of university education. The section that follows will explore further these various generative dimensions.

## Three dimensions of human enquiry

The pressing questions of human existence can be divided into three types: ontological – those relating to being, the nature of the self and the other entities that make up the world we live in; epistemological – those relating to knowledge, how we acquire it and what makes it valid; and axiological – the values that underpin what we do in our lives, the good and the just. These three philosophical categories may not encompass all areas of human enquiry, but they do represent the most weighty questions facing us. However, we do not need to see these categories as ones purely pertaining to the subject of philosophy. Instead, they are the bedrock of any educational endeavour. Whatever we set out to learn (and at whatever age or level of education), the process should engage us in reflection on questions of being, knowledge and value – as applied to the specific subject or content.

Climate change in this way has ontological, epistemological and axiological elements. It alerts us to questions of being (who we are and how we relate to the natural world), of knowledge (whose version of events should we trust) and value (what might be a fair distribution of the burdens of change). What this chapter will argue is that climate change is a particularly conducive issue for fostering deep reflection and for transforming learners. It serves not only its own agenda, of equipping learners to act in the climate crisis, but is a driver for a more powerful and transformative educational experience overall.

The sections that follow will address these three spheres of human enquiry in turn, drawing out the aspects of relevance to climate change, and the ways in which they can be explored in the classroom. In the space available, the discussion will highlight just a few ways in which climate change can be viewed through these three perspectives – undoubtedly there are many others.

## Ontological

The first, most immediate ontological concern is with the nature of the climate system, and by extension the nature of the world we live in. The temperature on Earth and its weather patterns are the result of an intricate interplay of factors involving the sun, the atmosphere, land, oceans and ice sheets, as well as living organisms. The consequences of temperature rises are also multiple (for example, melting of permafrost, changes to ocean currents), and many of them in turn influence the temperatures themselves, leading to feedback loops within the cycle. These loops create the possibility of tipping points, and the risk of no return from temperature rises (Lenton et al. 2019). The key ontological idea that emerges here is interdependence, and through a study of these processes, learners can reflect not only on the concrete and practical considerations of the causes of climate change and possible interventions, but also the deeper interconnectedness of the natural environment. In order to operate in the world, we designate categories and attach terms to them (for example, cloud, carbon, plants) yet when we reflect on their existence, it is clear that they are not entirely separate from other phenomena.

There is a long tradition of thought about interdependence in ecology – for example, James Lovelock's (1972) Gaia hypothesis, Lynn Margulis's (1998) symbiosis, Fritjof Capra's (1996) web of life and Anna Tsing's (2015) assemblages. While these accounts have been challenged and reframed, and should not be presented as doctrine, they serve a vital role in disturbing our conventional notions of the separateness and independence of phenomena, and allow new understandings to emerge. Many argue that this interconnectedness requires new forms of thinking: Lehtonen et al. (2019), for example, propose 'phenomena-based' learning in place of fragmented disciplinary divisions, through which wicked problems can be approached with systemic and holistic thinking.

The second area of ontological reflection is the human being. Climate change also challenges our conceptions of who we are individually and collectively. These reflections are in part similar to those above relating to interdependence. We have a physical body that occupies its own space, but on reflection we soon see how even our bodies are not really 'ours', made up as we are of millions of bacteria, and of water that is constantly recycled. Our mental world is also constituted by interactions with others and collectively constructed languages and concepts. The distancing of the human being from the natural environment in early modern Europe, and Francis Bacon's exhortation to human dominion over nature, can thus be seen not only as damaging but also based on a misconception (Merchant 1980; Silova 2021).[2] These questions can be fruitfully explored in the university curriculum, with arts, humanities, media and cultural studies having an important role in analysing the changing nature of human beings and human societies and the implications for the climate crisis (for example, Hawkey 2023). Studies of literature in universities might be considered an area in which it is difficult to integrate climate change, but analysis of artistic works can be a highly generative means of developing understanding of framings of the self as separated from 'nature' and from human communities. This analysis may involve authors who explicitly deal with the topic of climate change – for example, Amitav Ghosh or works from the 'cli-fi' genre (such as Octavia Butler, Margaret Atwood, Kim Stanley Robinson and Paolo Bacigalupi) – but also those that may predate awareness of anthropogenic impacts on the climate but nevertheless address crucial aspects of the relationship with the natural world that brought humanity to this point.

As part of the pedagogical process of reflecting on ontology, it is useful to consider alternative ontologies to those many students will be accustomed to (Komatsu et al. 2021). In a Western context, ideas of *sumak kawsay* (or *buen vivir*, good living) from the Andean region of South America and *ubuntu* from Southern Africa, which present a worldview of interconnected existence, are helpful in opening up the imagination (Assié-Lumumba 2017; Brown and McCowan 2018; Olivera Rodríguez 2017).

A third series of ontological reflections concern the notion of change itself. What is the nature of change, and what implications does its inevitability have for our understandings of space and time? Mortality, the fragility of human existence, the possibility of the end of the human species and continuation of planetary life without human beings, are all brought into the foreground. Naturally, some of these are disturbing topics and have to be dealt with in delicate ways, particularly in the context of widespread anxiety among young people (Hickman et al. 2021; Ojala 2016). Nevertheless, a careful treatment can be an important part of the deepening understanding of self and the future of humanity.

### Epistemological

Epistemic concerns are as old as conscious human beings but have taken on particular complexity in the contemporary age. These shifts can be attributed variously to: the coexistence of strongly contrasting epistemologies and ontologies within and between societies; the development of information and communications technology that brings individuals into closer contact with that diversity; the ready availability of huge stores of information through the internet, but without a clear criterion of sorting and selection; and the deliberate use of social media to spread misinformation and manipulate audiences.

Climate change provides a focal point through which all of these contemporary trends can be brought out into the open and reflected on in the classroom. It is an issue that is subject to significant contestation, in its most extreme form involving outright denial of its existence, but also a range of reasoned positions in terms of how best to address the challenge. Understanding and navigating this contestation is essential, not only in maintaining some degree of social cohesion and cooperation, but also in finding valid solutions – given the highly complex nature of the climate emergency.

As discussed in Chapter 3, climate change deniers have successfully been able to exploit the general value of scepticism in science, claiming that the environmentalist lobby is shutting down debate and stifling dissenting views, utilising the peer review system and other outlets of communication for their ends (Morano 2018). The academic community is then forced into either an unenviable position of asserting a uniformity of view – an anathema to science generally, and not entirely true, since even among those in agreement with anthropogenic climate change there are some differences of position – or one of climate science as an open question, which lays open cracks into which the denial lobby can hammer their wedges (Marshall 2014).

Many of these debates have involved the scientific consensus on climate change. Deniers have paraded the small number of scientists who contest anthropogenic global warming and used them to challenge the existence of a consensus. However, a bibliographic analysis (Cook et al. 2013) showed that 97 per cent of published articles with a position on climate change endorsed human-caused global warming. In order to navigate these contradictory messages, therefore, scientific literacy in relation to climate change is essential for all students of higher education and the general public. Naturally, it is impossible for all people to have a comprehensive knowledge of climate science, or the technical expertise of climate scientists, but a basic understanding is essential for underpinning one's own actions and in navigating the epistemic conflict around climate change. Importantly, however, this kind of awareness is not only one of separating the true from the false, but in developing a more nuanced awareness of different epistemologies that might provide multiple valid readings.

The role of higher education teachers, then, in relation to climate change is both to provide a space for learners to explore their own epistemological positions, individually and in relation to one another, and to present the contested epistemic terrain on which battles around climate change are being fought. Students, for example, could be asked to analyse a television debate between a climate scientist and a fossil fuel lobbyist, assessing both the rhetorical devices and communication approaches used by each, and the factual basis and reasoning employed in their arguments. They could then be encouraged to reflect on their own positionings in relation to the contrasting positions. These contestations involve an intertwining of questions of fact and value, of the contrasting logics of different disciplinary areas, of different epistemological and ontological foundations, and of complex systems (societal and environmental) without predictable linear outcomes.

Climate change also raises awareness of academic disciplines, including the relationships between them, and their contributions and limitations. In no sense can the phenomenon of climate change be limited to a single discipline: while certain disciplines may have a key role in identifying changes in climate and their immediate impacts (geology, geography, meteorology etc), the wide-ranging disruption to plant and animal life, human societies and all aspects of the natural and physical world necessarily means that all disciplines are involved. Economics, for example, now involves treatment of the question by default, including calculating the economic impacts of different temperature levels, building environmental damage into cost–benefit calculations, and assessing 'discounting' (the perception of decreasing value of benefits that occur in the future). The complexity of the topic makes it essential to include contributions from different disciplines, but also to make sure that disciplines work together in new ways, in transdisciplinary as well as multidisciplinary modes.

Another important dimension of epistemological awareness relates to the diversity of knowledge traditions. Human cultures have generated a variety of worldviews, which involve distinctive ontologies, epistemologies and axiologies. To acknowledge the value of understanding and engaging with these diverse knowledge traditions is not to slip into a limp relativism: different traditions may have their strengths and weaknesses, their areas of contribution and their applicability to different contexts and situations, or simply reveal different dimensions of human experience. While a problematic term in its own right, what we think of as 'Western' knowledge dominates higher education in today's world – through its formation of the inductive scientific method that provides the gold standard in our epistemic space, and through the institution of the European university, which has spread through the world. Western academic knowledge has many positive qualities and has brought undeniable achievements but it does not have an exclusive claim to truth or value; furthermore, Western traditions are themselves plural, and even practice in the natural sciences involves a variety of epistemic approaches, including intuition, imagination and tacit knowledge (Polanyi 2009; Binagwaho et al. 2022).

In order to safeguard this epistemic pluralism, and as a question of justice for colonised and subjugated peoples worldwide, there have been worldwide calls for decolonisation of higher education, spurred on by the prominent Rhodes Must Fall movement in South Africa (Del Monte and Posholi 2021). Indigenous movements have called for the inclusion of their knowledge traditions, not instead of but alongside mainstream ones, in line with an 'ecology of knowledges' (Santos 2015).

Given the complexity of the ecological challenge, and the forms of thought that led us into trouble in the first place, it is not unreasonable to believe that an ecology of knowledges will in fact be necessary to address climate change. Dialogue is needed between universities and external communities, between mainstream academic knowledge and local knowledge systems, and between different cultures and positionalities. Co-construction of knowledge becomes therefore a question of justice (so as to include participating communities as *subjects* rather than *objects* of the process of knowledge production and application), but also the most effective form of building lasting responses to the climate crisis. More fundamentally, many have argued that a whole new paradigm for humanity is needed (Silova 2021), to haul ourselves out of the pit created for us by millennia of exploitative relations with the non-human environment, made catastrophic by the increase in technological power since the Industrial Revolution, and through the increase of incentives for accumulation resulting from the growth of capitalism. Different visions of this relationship can be found in philosophies such as *sumak kawsay* or *ubuntu* discussed above, but also within marginalised Western traditions, such as eco-feminism and deep ecology (Assié-Lumumba 2017; Brown and McCowan 2018; King 1995; Olivera Rodríguez 2017; Sessions 1987).

## Axiological

Climate change is contested not only in terms of its sources of evidence and paradigms of understanding, but also in the value sets that accompany them. In fact, many (for example, Marshall 2014; Norgaard 2011) have argued that climate change denial is better understood as a conflict of values and emotions than a dispute over facts. The value contestations relate to various areas – the good life and forms of living that are seen to be worthwhile, questions of justice, what a fair distribution of the burden of change and disruption should be – as well as to questions of authority and freedom, the legitimacy of coercion, and what forms of organisation are necessary for achieving global sustainability.

There is, at the present moment, an indisputable situation of climate injustice in which the wealthiest communities and countries as a general rule bear disproportionate responsibility for causing climate change, while the poorest communities and countries bear the brunt of the negative impacts and lack the financial resources to protect themselves from them (Hickel 2021; Klein 2014, 2019). While mitigation (preventing the root causes of climate change), adaptation (adjusting to the new conditions) and regeneration (creating new forms of sustainable community and environment) are needed globally, the burdens of adaptation and regeneration on some are caused by the lack of attention to mitigation on the part of others. This is a geographical, political and economic divide (designated imperfectly by the Global North/Global South labels) but it also highlights inequalities within countries – leading to a focus on MAPA (the Most Affected People and Areas), cutting across different locations. These injustices have an intersectional dimension (Crenshaw 1991), in that social identities such as gender and race can compound the economic and social class disadvantage and create configurations that cannot be addressed in isolation.

While the facts of the case are clear, the implications are hotly contested. Do wealthy regions have responsibility for atoning for historical actions (for example, the Industrial Revolution in Britain) or only their current emissions? Should the largest emitters of greenhouse gases make financial payments to compensate other regions affected by the impacts? Should low-income countries be inhibited from developing fossil fuel-based industry when other regions of the world have historically generated their wealth from them? Addressing these questions head on in pedagogical spaces is important both for ensuring all are aware of the injustices, but also in refining learners' abilities to reflect, deliberate and position themselves on these complex issues.

Important value questions are also raised over the forms of social organisation necessary and permissible. Much of the opposition to environmentalism has been provoked by the constraints that it is seen to impose on individual freedom: that the movement is 'green on the outside, red on the inside' (Delingpole 2012), essentially communism by the back door, or else big government or even global rule by the United Nations. While these concerns are wildly inflated (and many environmentalists are equally concerned about constraints on individual freedoms and the dangers of excessive state power), resolving the climate crisis may indeed require limitations on individuals and corporations, and the establishment of new forms of global coordination (Helm 2020). If people do not make the necessary changes of their own volition, should they be forced to do so, and at what proximity to the precipice of species destruction would authoritarian measures be justified?

Finally, there are questions about the good life. Is our task as humanity to maximise (through technology and economic management) the possibilities of continuing the high consumption lifestyle that those in privileged parts of the world have become accustomed to? Or is a more frugal and less wasteful lifestyle, closer to nature and valuing the spiritual over the material, in fact a richer life in any event? Major religions have had ambiguous relationships with climate action, with Christianity on the one hand being held responsible for the root cause of the crisis in positioning the human being as the 'master' of nature, but on the other hand, as seen in Pope Francis's (2015) *Laudato Si'*, advocating for major pro-environment shifts. Climate change challenges us and causes us to question all aspects of the values with which we live.

Engaging with these values is a complex matter. Transmitting a predefined set of values to learners is challenging with young children, but almost impossible with adults of university age – and of dubious legitimacy even if it were feasible. What universities can do is allow space for learners to grapple with these complex questions head on, to appreciate their underpinning principles, to understand diverse positions and expand their moral reasoning to those in different positions, places and points in time. Subjects in the arts, humanities and social sciences will engage more readily with these value contestations, with the teaching of history, for example, having an important role in generating understanding of the role of empire, slavery and capitalism in the progressive destruction of natural environments (Hawkey 2023). Yet it is important that such value considerations are brought into STEM subject areas as well, as they represent crucial ethical and civic issues that must be at the centre of science teaching in universities and beyond (Salinas et al. 2022; Torres-Olave and Bravo González 2021). For example, discussions of geo-engineering and carbon capture should involve not only the technical aspects, but also questions of justice surrounding who owns and controls the technology, and the impacts on diverse communities.

This section on the three dimensions has argued that climate change can act as a positive driver for change in teaching and learning, in opening up the profound ontological, epistemological and axiological questions that all education should address. It is true that any issue one could choose to study (from the ancient Greeks to US–China trade relations, genetically modified crops to quantum computing) could potentially be addressed from these three different angles. But climate change is particularly conducive to opening up crucial questions and dilemmas in these areas, through its complexity, its moral urgency, its comprehensiveness (in touching on all aspects of human existence) and its global reach (in involving all of humanity).

## Pedagogical foundations

The above sections have set out three broad areas of enquiry in relation to climate change, ones which provoke deep questions about ourselves and the world, and are conducive to the transformation of self and society. Yet there are a range of possible ways in which these questions can be addressed in the classroom. Some attention, therefore, is needed to the *process* elements, to the orientations of teaching and learning in the classroom (Alexander 2010). This section will not outline specific teaching techniques or resources, but instead highlight two fundamental principles – critical questioning and deliberation.

Given the urgency of the issue, and the high degree of scientific agreement, it might be tempting to present climate change as a settled set of facts, commitments and actions to be instilled in students. Yet, as argued by Jickling and Wals (2008) in relation to education for sustainable development, a campaigning or advocacy approach is never justifiable in the classroom. In an educational setting, particularly one involving adults, learners must exercise their own agency to engage with the material and acquire new understandings through processes of critical reflection. Research has, in fact, shown the dubious efficacy of awareness-raising of a purely cognitive nature (Anderson, A. 2012; Bangay and Blum 2010; Facer 2020; Facer et al. 2020; Monroe et al. 2019; Oberman and Sainz 2021; Rousell and Cutter-Mackenzie-Knowles 2020; Stevenson et al. 2017). In the case of climate change, the complexity of the issue means that creativity and imagination must constantly be employed to adapt to the emergent properties of the system and form new responses to the crisis. Conditioning, non-reflexive training or even subliminal messages might be successful in bringing about pro-environmental behaviours in the short term. But they are not solutions to the climate emergency.

Processes are needed, therefore, through which learners can become aware of their own understandings and positions, engage with other perspectives and worldviews, challenge their assumptions and construct new possibilities. The dual processes of critical questioning and deliberation are central here. These two principles are fundamental to the educational process, as they encourage reflection, perspective and possible revision of our views and understandings. As principles they are applicable to any educational setting and any subject matter, although they are far from straightforward to implement and can be challenging, as they disturb the comfort of our familiar and entrenched ideas.

Questioning and deliberation are approaches that we as educators bring to the teaching of climate change with a normative orientation (Freire 1970; Gutmann 1987; Nussbaum 1997). They emerge from commitments to human agency, respect for persons and the value of human understanding, rather than unreflective survival or subordination of the human being to external goals and technologies. Nevertheless, these prior commitments require conducive subject matter in order to be operationalised in an educational setting. The complexity, profundity and contestation around climate lend themselves both to processes of critical questioning and to vibrant group discussions, leading to a virtuous cycle of mutual reinforcement through which the principles can be deepened.

## Critical questioning

Educational spaces can be structured so as to encourage learners to question their existing beliefs, perspectives and assumptions. This process operates on a continuum, from relatively mild revision of one's factual knowledge, to a fundamental about-turn in one's identity and worldview. Critical questioning is opposed to learning approaches that are transmissive, involving an unquestioning flow of knowledge from teacher to student, and also to learner-led processes that involve accommodation of new knowledge entirely within existing assumptions (Browne and Freeman 2000; Kuhn 1999; Lipman 1988).

While promotion of critical questioning is largely a matter of underlying orientation of the teacher, and can manifest itself in multiple ways, there are some recognised formal approaches. The Socratic method is the root of many of these approaches in the Western tradition. Socrates aimed to spark insights in his interlocutors by taking them through a series of searching questions that would force them to reassess their unexamined assumptions. This generation of new knowledge through questioning has been an ever-present current in the Western higher education tradition, alongside traditions of transmission of knowledge and memorisation (Barnett 1997). Socrates described himself as a 'gadfly', niggling at and disturbing the complacency of Athens. The idea of this form of questioning being uncomfortable is a common theme in critical approaches (for example, Sterling 2011). While challenges to and reframing of our fundamental assumptions are ultimately beneficial, they are unsettling and at times painful.

A more recent structured approach is problem-based learning. Instead of approaching learning from the starting point of a body of knowledge to be acquired or mastered, it starts with the solving of problems relevant to the professional area – either theoretically contrived ones or ones encountered in actual practice (Schendel et al. 2023; Williams 2001). Problem-based learning is commonly used in universities, being particularly prominent in health sciences, in which trainee doctors and nurses develop not through acquiring a formal body of knowledge but through being faced with real-life medical situations.

Climate change is embedded in human civilisation, practices and belief systems and so addressing it involves critical questioning of this sort. The transmission of a body of knowledge relating to climate science is not entirely worthless – certainly there is some factual knowledge that all people should have – but it is unlikely to be sufficient for finding solutions to 'wicked' problems or to bring about the kinds of individual and collective changes that are necessary for a sustainable planet. These various forms, such as Socratic questioning and problem-based learning, are essential for, in the first place, sparking realisations about the complex web of causes of our current unsustainable lives and societies, and then thinking creatively about how to move forward. These approaches can be adopted in the various one-on-one teaching situations in the university (for example, in postgraduate research supervision or in tutoring for an essay or dissertation at undergraduate level), in which Socratic dialogue is readily applicable. Group situations are conducive to problem-based learning, but forms of Socratic questioning through discussion are also possible, along with other methods such as simulations, role plays and thought experiments.

Further to the above, there is a tradition of more political questioning, focusing not so much on challenging assumptions of our identity and existence or solving problems, but of challenging and overcoming the injustices that exist in our societies. Most prominent of the thinkers associated with this current is Paulo Freire (for example, 1970), whose primary insight was that education inevitably serves a political purpose – in his terms, either liberating or domesticating. This influence is not so much because of the explicit content – although in some cases there will be direct treatment of political issues in the classroom – but because of a deeper process of formation of the 'subject' or person. Freire observed a correspondence between the disempowerment of the learner – considered to be an empty vessel, with her existing learning and knowledge disregarded – and the disempowerment of the citizen in the political sphere.

In Freire's pedagogy, questioning occurred initially through the presentation of visual cues (stylised representations of the present reality), intended to provoke reflection on learners' conditions of living and inequalities in society. More broadly, Freire (1970; 1994) advocated for problematisation or problem-posing education, through which the naturalisation of disparities of power and wealth could be challenged. Problematisation is practised in the educational space hand in hand with dialogue – used by Freire in the sense of respectful, horizontal pedagogical relations – which together lay the foundations for individuals taking the reins of their own destinies in the broader world. This complementarity is similar to that between questioning and deliberation discussed elsewhere in this section.

In his own writing, Freire's main concern was poverty and oppression, and not the natural environment, although he was said to be writing a book about the latter at the time of his death (Misiaszek 2020a). However, it has since become clear that environmental concerns are no longer those of the privileged middle class with leisure time to enjoy nature, but intimately bound up with global social justice and the well-being and survival of the poorest communities (Klein 2014; 2019). Freirean conscientisation in the twenty-first century inevitably involves a critical understanding of climate change and its causes, and coordinated collective action to address it. The kind of action that will emerge from transformative pedagogy is not the isolated, top-down, technical solution of geo-engineering or carbon capture, but a transformation of our local, national and global economic, political and cultural systems to put in place a more caring, egalitarian and sustainable world. These transformations involve not only Freire's initial concerns for working-class oppression, but also ones relating to gender, race, coloniality and other forms of social identity.

Freirean approaches of problematisation and conscientisation are, therefore, highly relevant to climate. Collective processes of analysis and reflection can reveal the ways in which climate change is bound up with socio-economic inequalities and asymmetries of power at all levels, and how the solutions need to be grounded in the fostering of more egalitarian and just societies. Climate change represents a teaching opportunity in this sense: through sustained analysis and reflection, what initially appears to be a neutral technical issue reveals its roots in distribution of resources, modes of political decision-making, power differentials and our entire civilisational model. So, for example, practices with wide acceptance in society, such as recycling and green consumer choices, can be subjected to critical scrutiny, highlighting their roots in the fossil fuel lobby's deflection from needed structural changes towards individual responsibility.

While there are those (for example, Bowers and Apffel-Marglin 2005) who argue that the Western anthropocentric currents underpinning Freire's thought are inimical to ecological sustainability, his thought has been integrated with environmental ideas through the eco-pedagogy movement (Gadotti 2000; Kahn 2010; Misiaszek 2020a, 2020b). Eco-pedagogy represents the educational manifestation of the uniting of the social justice and ecological agendas; as Jacobi (2003: 189) states, environmental education must be 'above all a political act oriented towards social transformation'.[3] Critical questioning, in this way, runs a full arc from more technical approaches to rational argument, to more political processes aiming for a fundamental transformation of society.

## Deliberation

A fundamental part of living in a collectivity is deliberation – at least if we are to avoid authoritarian or absolute rule. Listening to the views of others, communicating our own views, and then, through the interaction of the two, revising our views, are essential both for making the right decisions and for ensuring justice and inclusion in society. As argued by many commentators over the years (for example, Gutmann 1987; Mill 1991 [1861]; Pateman 1970), deliberation is not only a guard against authoritarian rule, but also against forms of majoritarian democracy that reduce the democratic principle to a competition of rigid positions.

Higher education is a highly conducive space for the development of deliberation – a practice that must be learned through experience. The full possibilities of deliberation depend in the first instance on policies of access and ensuring that university spaces do not become segregated on the basis of socio-economic background or other factors. But they also depend on the pedagogical environment created in the classroom. Many developments in teaching and learning in higher education over recent decades have in fact focused on creating a space for deliberation in the classroom, by fostering an environment in which students feel able to raise questions and by protecting the time available for these discussions – particularly through flipped or inverted classroom approaches where the content input takes place largely before the real-time class (Lage et al. 2000). Deliberation can take place in online fora, as an alternative to face-to-face environments, but careful consideration is needed to ensure that the design of the virtual space allows for these forms of interaction.

Deliberation in all spheres involves dealing with disagreements, some of which are sensitive and heated, and relate to value-based questions without clear answers. In higher education, these controversial issues are a challenge but also an opportunity: a challenge because they are hard to present and frame on the part of the teacher, and because they can fuel tensions and conflict among students in the classroom; but an opportunity because their charged and ambiguous nature means that they can provide intense engagement in the educational space, foster critical dialogue across diversity, and expand students' nuanced moral reasoning and action. While levels of concern about climate change are generally high among youth populations (Hickman et al. 2021), and (depending on the context) the proportion of those denying climate change outright is likely to be low, there will still be significant differences in students' views on how best to address the challenge – in line with the axiological divergences outlined above.

In deciding on how to incorporate deliberation into the teaching of climate change, there are certainly arguments in favour of excluding climate denial. First, the evidence and scientific research available to us at the present moment shows that anthropogenic global warming is real (Masson-Delmotte et al. 2021), so any fundamental challenge to that view could be prohibited on the basis of spreading false information. Second, given the Herculean task of transforming an unsustainable society into a sustainable one, and the catastrophic costs of not doing so, allowing voices to undermine that task might be considered too great a risk to take.

On the other hand, allowing climate change denial in the classroom enables exploration of the contested epistemic dimensions of climate change, as discussed above: these could be explored theoretically, but may be more vivid if embodied in the views of participants. Another reason is that the exclusion of climate denial puts it underground, which paradoxically allows it to survive and even flourish, as a consciously countercultural view. That said, there may be versions of climate change denial that would, in any circumstances, be inappropriate in the classroom – particularly if linked with racist, sexist, homophobic or other exclusionary and prejudiced views and incitements.

Monroe et al.'s (2019) systematic review showed the value of '*deliberative discussion* to help learners better understand their own and others' viewpoints and knowledge about climate change' (p.11, original emphasis). Yet creating this kind of environment is not straightforward, and providing a real space for deliberation in classrooms can at times be threatening for higher education teachers: it involves letting go and allowing the learners to dictate the movement of the discussion, and risks disagreement and even conflict. For learners too, it may be an uncomfortable experience. But, as argued by Kwauk and Casey (2021), there is value in the disruption of ideas, and even in making learners intellectually uncomfortable as a way of unsettling entrenched and unquestioned views – and contested issues can be useful for this end.

Critical questioning and deliberation are not new ideas and have been seen in educational traditions around the world from the monasteries of classical India to Scandinavian folk high schools (Ellis 2019). Yet while ever present, they are always vulnerable to the convenience of the standardised, transmission approaches to teaching, ones which we associate with either marketised or authoritarian education systems, but which can equally be a temptation when faced with an urgent moral issue such as climate change. Critical questioning and deliberation underpin the treatment of the ontological, epistemological and axiological subject matter in the classroom, providing a foundational orientation from which the specific methods and approaches to teaching and learning can emerge. These methods can involve not only group discussion in the classroom, but also arts-based approaches, role plays and simulations, storytelling and many others.

These two fundamental principles are not separate but interact and are complimentary. Questioning is, in the first instance, an internal process – subjecting to critical scrutiny the assumptions held by the individual – although it will often occur between teacher and student or in a group situation. Deliberation occurs primarily through engagement with others, although it can also occur internally, through the process of self-reflection. But deliberation with others is a key means of fostering questioning, and critical questioning is an ever-present part of deliberation. Critical questioning and deliberation, therefore, have their own intrinsic value, but in practice occur in conjunction, with each enabling and strengthening the other. The ways in which they manifest themselves in real-life education depend naturally on the context and circumstances, and they can appear in a multiplicity of forms while still adhering to the foundational principles.

## Towards pedagogical renewal

Misiaszek and Rodrigues (2022: 5) call for a 'paradigm shift in HE [higher education] to much more directly focus on teaching to achieve JBSE [justice-based sustainability education] globally' and for 'thorough and meaningful transdisciplinary incorporation of JBES throughout all HE curricula'. Yet, while we need an 'epistemological paradigm-shift for most HEIs, including reinventions of disciplines and their epistemological foundations', this shift is unlikely to take place overnight, at least not in mainstream institutions. Moving towards this aim requires provisionally working within established courses of a disciplinary or professional nature in the meantime – not conforming to them, but challenging them and pushing their boundaries from within.

This renewal is urgently needed in higher education. While the sector has seen startling growth in recent decades, its positioning as a mechanism for labour market allocation has stored up major problems, being unable to fulfil all students' aspirations for social mobility and simultaneously being distracted from its foundational role of providing a space for intellectual exploration and transformation. As highlighted by Stein et al. (2023) and Jimenez and Kabachnik (2023), it is implicated in the continuing mindset of 'progress' based on extraction and exploitation, with HEIs' endorsement of sustainability goals very often being rhetorical or grounded in an illusory 'green growth'. Higher education must become (or return to being) a deeply transformative experience, leading us to engage with the most profound questions of our being, knowledge and values, avoiding monocultures of the mind (Shiva 1993) and instead developing critical deliberative environments and an ecology of knowledges.

Fortunately, despite neoliberal designs on higher education, universities have maintained some spaces of autonomy from the market, and institutional traditions provide some protections for counter-hegemonic work. As argued by Binagwaho et al. (2022), even the Western scientific method is founded on scepticism of monolithic bodies of knowledge – as shown in the British Royal Society motto of *nullius in verba* ('take nobody's word for it') – so it should, in theory at least, leave the door open for epistemic pluralism. In recent years, decolonial movements in the Global South and internationalised student bodies in the Global North have also opened up a space for challenging conventional curricula and creating possibilities for an ecology of knowledges (Del Monte and Posholi 2021).

As discussed above, one of Freire's (1970; 1994) most powerful insights was into the inevitably political nature of pedagogical interactions – not only because they often directly deal with political content, but because they involve the formation of agents, leading either to disempowerment or empowerment of the learner and citizen. The implication here is that teachers cannot 'sit on the fence' and remain neutral in their teaching: they are either liberating or domesticating. In the same way, teachers can (no longer) avoid including climate change in their teaching. Its centrality to the fate of humanity means that it is inevitably part of any meaningful discussion of society and the natural environment, and so part of every disciplinary and professional area. Not addressing it means supporting the current slide into self-destruction for humanity. Given the questions of environmental justice alluded to above, not addressing climate change also means perpetuating inequalities at all levels.

This chapter has explored these ideas in relation to the transformative learning that is essential for responding to the climate emergency. Three spheres of human enquiry have been highlighted – the ontological, the epistemological and the axiological – underpinned by critical questioning and deliberation. In each case, climate change can be seen to represent a stimulus to change – a challenging and unsettling one, but one that can bring a much-needed shift. While effective teaching of climate change depends on a pedagogical approach oriented around various factors (an open classroom environment for discussion, building on students' existing knowledge, experiential learning, acknowledging emotions, use of arts, engaging with activism and so on; Ojala 2016; Bryan 2020; Lehtonen et al. 2019; Nussey 2021), it also in turn stimulates these active pedagogies in a virtuous cycle. While this chapter has focused on those parts of students' learning experience that are controlled by lecturers, it must be recalled that there are many other aspects (perhaps equally important), including peer learning and self-directed learning outside the classroom. These broader spaces of learning will be addressed in the chapter that follows.

## Notes

1 An approach in which, instead of class time being used to transmit knowledge content (for example, in the form of a lecture) which students then apply afterwards, students acquire the knowledge in advance through reading, videos and so on, leaving the class for questions, discussion and application of the material.

2 This is not a comment on the moral and political desirability of individualism versus collectivism, but an ontological point about the extent of separateness of human beings.

3 Translation from the original Portuguese by the author.

# 7
# Curriculum topography

Calls for greater attention to climate change in the curricula of schools and universities around the world have been very welcome. If dealt with at all, climate change has usually been included in natural sciences and geography, and the important task of integrating it into other disciplinary areas requires some unsettling of conventional course content (Hess and Collins 2018; Leal Filho 2010; Nugent 2021; Reimers 2021; Rousell and Cutter-Mackenzie-Knowles 2020). Yet, while carving out greater space for climate change within the formal curriculum is undoubtedly important, it is just part of how climate change can be integrated into the educational experience.

This chapter puts forward a vision for the role of education built around the idea of a curriculum 'topography'. This approach sees the school or university as an educational environment with a diverse range of opportunities for learning – taught, self-directed and through peer collaboration – in distinct spaces within and beyond the physical campus and virtual space of the institution. The task of the educational institution is to curate this curriculum topography to provide the richest possible learning environment, in which all students can develop their understanding of, engagement with and action in response to climate change.

Universities and other HEIs have some particular characteristics that mark them out as different from compulsory school-level provision for children. These include their institutional make-up, with a variety of functions such as research and public engagement outside of the educational offering, and learners who are predominantly adults, with high levels of choice over their studies. The availability of research on curriculum relating to climate change in higher education is growing. Yet, for the most part, these are analyses of specific courses (for example, Amos and Carvalho 2020; Fahey 2012), or of the integration across the taught curriculum (for example, Hess and Collins 2018; Gomes 2020), with some studies of teaching and learning approaches in formal and informal spaces (for example, Bush et al. 2017; Rooney and McMillin 2010; Senbel et al. 2014). This chapter takes a broader approach in assessing the whole of the learning environment, within the university and beyond, seeking to understand the relationships between the diverse spaces and the structural features underlying them.

The curriculum is here understood as encompassing opportunities for learning that are organised by and through the university, involving not only formal taught courses but also interactions on the university campus, and voluntary and professional work outside whether lecturer-led or student-led. In discussing curriculum, it is always important to bear in mind the gaps that may exist between the stated intentions and pronouncements (the official curriculum) and what ends up being carried out in practice (the taught or unofficial curriculum), not to mention the submerged rituals that may not be apparent even to those engaged in them (the hidden curriculum). Across the various levels of education, the vast majority (95 per cent) of the 194 countries reporting to the UNFCCC state that they offer some climate change education in their curricula (UNESCO 2019), yet there is little evidence that this official integration filters into the classroom. PISA results[1] show poor learning outcomes relating to the environment even in the relatively privileged OECD countries (OECD 2012).

There are some aspects of the topic that cannot be dealt with in full in this chapter. First, there is not space here to specify the exact content that will go into climate change courses (whether palaeoclimatology, geo-engineering or wind energy). Second, the chapter will not deal comprehensively with questions of pedagogy, teaching methods or learning styles, but will build on the broad discussions of the previous chapter. Third, while there is discussion of accreditation, there will not be a detailed treatment of assessment, although it is acknowledged that curriculum in practice may be strongly determined by what is assessed and that constructive alignment is crucial. Fourth, there are complex questions in curriculum studies about the nature of knowledge and whether it is culturally specific or universal (White 2019; Young 2008; Santos 2015), linking in with movements for the decolonisation of higher education in recent years. These are relevant for climate change education, and will be touched on here, but are dealt with in greater depth in other publications. In addition, the analysis presented here will chart the broad contours of climate change in the university experience, but the ideas presented must, needless to say, be contextualised in the specific circumstances of each university.

## Is it the role of the university to teach about climate change?

Before outlining a curriculum framework for higher education, it is important to assess whether the kinds of climate learning outlined in Chapter 3 should indeed be acquired in the university, or alternatively at school, in the family or in broader society. Answering this question involves determining what kind of institution a university is and what its general purposes are, a task that on its own could require multiple volumes. Nevertheless, a few words here will be of use. Higher education is generally taken to refer to a level of study undertaken after basic education has been completed, and therefore normally in adulthood, providing substantial depth and sophistication of learning, usually in a specialist area. The institution of university, which now dominates higher education globally, has its origins in medieval Europe, although it is only one of a number of historical manifestations of HEIs around the world (Carpentier 2019; McCowan 2019; Perkin 2007). The university is distinct from some other forms of adult education, vocational education and apprenticeship in that its primary purpose is critical and open-ended enquiry (Collini 2012). While universities come in many different guises, and conduct research and community engagement and other functions to greater or lesser degrees, this purpose of critical enquiry would appear to be a unifying factor – their educational approach going beyond mere training of a non-reflexive type.

The characteristics outlined above can shed some light on the place of climate change within the institution. Looking back to the distinction between knowledge, skills and values drawn in Chapter 3, few would question whether it is the place of the universities to promote the first of these. University is the knowledge institution par excellence, both in terms of passing on knowledge to students and generating new knowledge through research and scholarship. There are, nevertheless, significant debates about what knowledge specifically should be made available to students, whether a broad curriculum such as one would expect at school, or a tightly focused area, or professional or vocational knowledge as opposed to general knowledge. Countries have addressed these questions in different ways in their higher education traditions, with the USA, for example, leaning towards a broader general curriculum (in many cases in the liberal arts) and European nations towards a narrower focus in a single discipline.

Skills are rather more controversial. Few would question the development of discipline-specific skills (such as preparation of specimens in biology) or generic academic skills (such as argumentation and critique). While there are calls from industry for work-ready graduates, there is scepticism from some quarters as to whether universities should be promoting non-academic skills, whether of a vocational nature or more general life skills (McCowan 2015b). It can be argued that these skills are better developed in other settings (such as in the workplace itself) or should have been developed at an earlier stage (in school or in the family).

Yet it is in the sphere of values that the most doubt exists. While it is likely that most of our fundamental values are already in place by the age that people commonly go to university, some shaping of values is inevitable, as it would be in any long-term experience. With the massification and in some contexts near universalisation of higher education, the civic role of the university becomes more prominent. When the majority of citizens in the country are attending a higher level of education it inevitably takes on an important role in shaping the interactions and practices of society as a whole, influencing its democratic or alternatively authoritarian character and the level and nature of political participation. The urgency and weightiness of climate change would seem to qualify it as one of those issues that cannot be ignored by an educational institution attending to a substantial proportion of the population.

Yet many argue that it is not the place of the university to promote a set of values – at least values beyond those of the academic discipline, or that justifying higher education on the basis of an overarching civic goal is unduly paternalistic (Martin 2022). Universities have been bastions of ethical individualism (Dworkin 1996) in which personal convictions are sacrosanct; it is against the principle of academic freedom to oblige all staff to adhere to a particular set of values – other than, perhaps, procedural values such as academic freedom itself. Even if it were not against the principles of the institution, it would be highly difficult in practice to ensure unity of values in such large, diverse and autonomous organisations (Haddock-Fraser et al. 2018). Furthermore, in any form of education, while creating space for reflection on and clarification of personal values can be very useful, it is perilous (and of dubious efficacy) to oblige teachers to instil specific values in students that they (teachers) may not hold themselves.

On the other hand, institutions can be seen as transmitting values even in the absence of an explicit or intentional effort to do so. As discussed in Chapter 8, sociology of education over the past half-century has highlighted the hidden curriculum of schools and universities: promoting conformity to the status quo and the acquisition of values and practices of competition, hierarchy and exclusion (Giroux and Penna 1979; Jackson 1968; Margolis 2001). By definition, these processes are hard to identify and control, and their outcomes are unpredictable, with students able to resist as well as absorb the messaging (Willis 1977).

Does the difficulty of ensuring a unified value set make impossible the task of climate change education? And does it put in doubt the advisability of aligning universities' work with frameworks such as the SDGs? Despite the thorny issue of values, there are still strong reasons why universities should incorporate climate change and sustainability into their curricula. Climate change (mitigation and adaptation) is so important for our survival, and is inevitably part of all people's existence for the foreseeable future, that we should include it in the university as part of preparation for life. Furthermore, climate change is now a crucial part of academic disciplines, in their attempts to describe and analyse the world around us, so it would be scientifically negligent to exclude this material. Equally, for professional preparation in many areas it is now essential to include capacities relating to climate action. Climate change education (or sustainability more broadly) can be pursued in universities, even in light of the constraints on value promotion, in ways that leave the aims and outcomes more open, on the basis of deliberation and enquiry, and are consistent with academic values.

The appropriate response of the institution committed to addressing climate change would therefore be to address it in those areas of action that it directly oversees (operations, cross-faculty initiatives, partnerships, external engagement), and to ensure an enabling environment for lecturers and students to pursue relevant learning and enquiry in the semi-autonomous spaces in which they operate and interact. It is this approach that will characterise the proposals below. Finally, it is important to emphasise that whatever responsibilities or possibilities higher education has for teaching climate change, it will never be the only important space in which people learn about it. The undergraduates participating in the four-country Climate-U survey, for example, put university in third place after internet/social media and mainstream media as a source of climate learning (Rolleston et al. 2023). The university needs to view its learning spaces in conjunction with those outside.

## Climate change in the formal curriculum

The most obvious way in which climate change can appear in the university curriculum is as a discrete course – whether an entire degree programme or a subcomponent of it. Full degree programmes do exist, many at the graduate level (for example, the Master of Climate Change Adaptation by Research at the University of the Sunshine Coast, Australia), but also some undergraduate programmes: in the UK, it is possible to study a full Bachelor of Science degree in Climate Change in universities such as Greenwich, Northampton and Liverpool John Moores. There are many more dual courses – for example, climate change and health, law or international development. These courses are crucial for those looking to forge a career in the various lines of environmental work. Yet few would go so far as to argue that all or most university students should study a dedicated degree in climate change.

More common is the provision of a unit or module on the topic, either as part of specific courses or freestanding. Examples of this form of provision are becoming more common, as are cases in which climate change appears as part of sustainable development or environmental education units. Sometimes courses are provided outside of the regular curriculum framework, such as the climate literacy course available to both students and staff at the University of Bath, certified by the Carbon Literacy Project.[2] Should this type of cross-cutting module be compulsory for all students, to guarantee coverage regardless of the course studied? This question is complex, for principled and pragmatic reasons. As discussed above, there are constraints on universities as institutions in compelling students to engage in particular activities, given that the learners in question are normally adults who have freely chosen to study there. From a pragmatic perspective, compulsory modules not directly connected with disciplinary content may not be taken seriously, or at worst provoke resistance on the part of students. Universities' dual roles as institutions of both learning (intrinsic and instrumental value) and accreditation (exchange value) are brought into tension here. Furthermore, there is pressure for space in curricula (with climate change competing with a number of other cross-cutting areas needing attention – peacebuilding, global health, human rights, citizenship and so on), and thought is needed as to how the content links in with other areas of study.

In any event, even if it were legitimate and advisable to compel students to engage in a module on climate change, it may not be the best approach educationally. A danger of the discrete subject approach – as seen in other areas such as citizenship – is that students then confine the topic to one part of their learning and fail to see the broader connections (McCowan 2009). Embedding climate change across all areas of the taught curriculum, while harder to achieve, is likely to be the more effective approach, alongside discrete modules made available on an optional basis.

A helpful framework in this regard is provided by Molthan-Hill et al. (2019). They distinguish between four approaches to embedding climate change in the curriculum: *piggybacking* – incorporation of content into existing modules; *specialising* – creation of new modules; *mainstreaming* – incorporation across the whole of the existing curriculum; and *connecting* – creating new cross-disciplinary offerings. The dynamics of interest here are the movements between the two axes of existing/new structures and narrow/broad curriculum. Pragmatically, an initial step for universities will usually be to piggyback onto existing structures, while simultaneously opening up spaces for the creation of new modules. Coverage is most straightforwardly provided in discrete modules, but ultimately incorporation across the whole curriculum is needed.

As illustrated in the Molthan-Hill et al. (2019) scheme, climate change education commonly mobilises arguments around interdisciplinarity. Silo working is seen to be one of the main barriers to successful practice in this area – both as regards sustainable development as a whole (hence greater attention to interlinkages between the constituent goals in the SDGs in comparison to the Millennium Development Goals [MDGs]) and in the university. The primary argument for interdisciplinary working is that the complexity and multifaceted nature of climate change means that solutions will only be possible if it is approached from different angles simultaneously. Moving from the research to the educational function of the university, it is argued that students should be nurtured within this broader interdisciplinary perspective, rather than the restricted view of individual disciplines (Facer 2020; McCowan et al. 2021; Binagwaho et al. 2022).

In clarifying this discussion, it is important to distinguish between multidisciplinarity, interdisciplinarity and transdisciplinarity, which can all be mobilised in support of the above aims, but which require increasingly deep levels of change. Multidisciplinary work involves researchers from different disciplines applying their disciplinary expertise to solve a common problem; the benefits of multiple perspectives are brought to bear on the problem but the disciplines remain intact through the process. When applied to teaching and learning, this process would entail separate sessions from different disciplinary perspectives on a common theme. Interdisciplinary work, on the other hand, brings the disciplines into dialogue with one another. The disciplines are still present but begin to modify each other as they highlight each other's distinctive characteristics, critiquing and revealing positive qualities. There is a unified final product – whether a research output or outcome of learning

– although drawing on the diverse disciplines. Finally, transdisciplinarity takes us beyond disciplines altogether. It is a new way of working that shows no barriers between conventional perspectives.

These distinctions help us to conceptualise the different ways in which climate change can appear in the curriculum. In its simplest form, climate change can appear on the syllabus for a regular disciplinary area – most obviously an area such as geography, geology or environmental science. A multidisciplinary perspective can be obtained from ensuring that climate change is embedded across many or even all courses (though in most cases it will not be possible for each student to access the multiple perspectives). An interdisciplinary approach would involve a unit being taken by students from various base disciplines, led by lecturers from a variety of areas, putting their different perspectives (and those of the students) into dialogue. An example of this kind of course is provided at the University of São Paulo, in the INCLINE centre, in which an interdisciplinary Master of Science module is provided for students from any disciplinary area, involving perspectives from palaeoclimatology, oceanography, economics, public policy and biometeorology. As argued by Fung (2017) and Burandt and Barth (2010), the interdisciplinary experience can occur as much through the mixing of students on different courses within the same classroom as through the combination of content from different disciplines. Transdisciplinary initiatives at universities are rare, since recruitment and promotion of staff usually follow disciplinary lines, and students normally slot into disciplinary or at least multidisciplinary courses. But an example would be a regular discussion group on climate change for students and staff, a space for learning that breaks with conventional epistemic divides and ensures engagement with the topic from fresh perspectives, organised on thematic rather than disciplinary lines.

There are a range of ways, therefore, in which climate change can be integrated into the formal curriculum, alongside other opportunities for learning on campus and beyond. But how can we understand these different possible locations of climate change in the curriculum, their relationships to one another, their key characteristics and implications for student learning? The framework put forward in the following section attempts to address these questions.

## A framework of curriculum topography in the university

Most approaches to climate change education take one of two approaches: either designing a discrete course, or determining a set of learning outcomes or competencies that will be achieved across a range of taught provision. These approaches more or less correspond to an input or output focus, respectively, by determining the kinds of treatment that students need, or the goals that need to be achieved. This chapter takes an approach that is different from both of these, one corresponding more to *process* than input or output (McCowan 2013). It outlines a *topography* of learning experiences with which students can engage. Topography is here used metaphorically to refer to the varied landscape of learning opportunities, with diverse opportunities for learning about climate change in different spaces, in different formats, from and with different people, and with different outcomes.

Drawing on the discussions in the previous sections, the curriculum topography approach is helpful because: it addresses the need for diverse forms of learning (knowledge, skills and values; learning about, for and through); it is based on learner agency rather than coercion; and it incorporates both discrete and embedded forms of learning of a

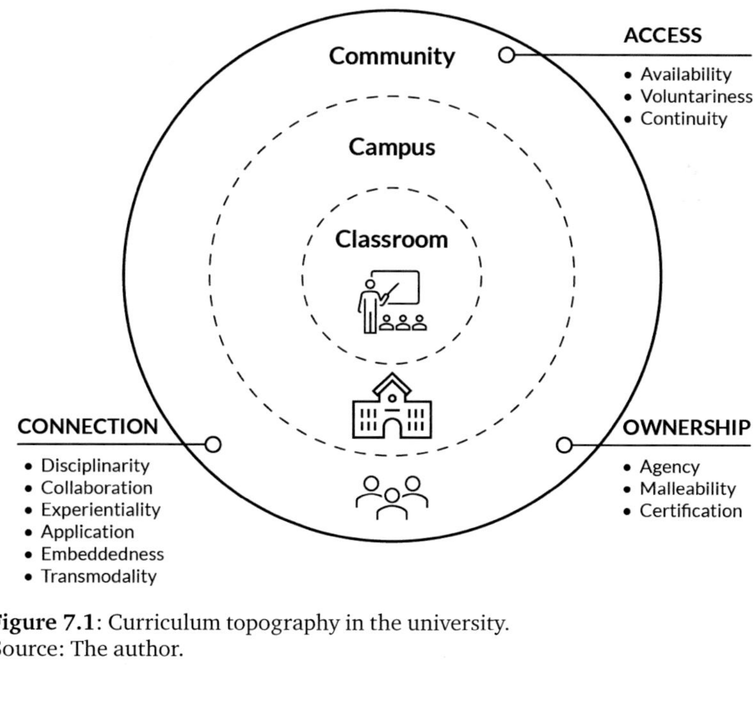


**Figure 7.1**: Curriculum topography in the university.
Source: The author.

disciplinary and non-disciplinary nature. In order to gauge the kinds of topography present in universities, and to plan for future transformations, we need to understand the spaces of learning that exist, and the relevant characteristics of those spaces. Figure 7.1 outlines these principal spaces and their features.

The graphic presents a framework for understanding the topography of learning about climate change in universities. It has two facets: first, the sites of learning (classroom, campus and community), the places where engagement takes place; and second, the curricular features (availability, voluntariness, continuity and so on) in three groups – access, ownership and connection. The curricular features can apply to forms of learning occurring in any of the three sites, although some are more likely to go together: for example, formal taught courses (classroom) are more likely to carry credit than engagement in a student society (campus) would.

Three primary sites of learning are designated by the 3 Cs of classroom, campus and community:[3]

- *Classroom:* The first C refers to the formal curriculum, which may or may not be delivered within a literal classroom. Provision here is likely to take the form of either discrete provision, such as an optional module on climate change, or being embedded across all modules, or both.
- *Campus:* Learning also takes place in other parts of the university space, whether physical or virtual. For example, through student associations, artistic and recreational activities, or through engagements in environmental initiatives on campus such as recycling or reducing energy usage.
- *Community:* Finally, learning can take place beyond the campus in activities organised by or through the university. These activities may be part of formal courses, such as research projects, linked internships or work placements. Or they may be entirely independent of students' programmes of study, such as volunteering work, campaigning and mobilisation. Community here is used in a figurative sense and may not literally be the local community surrounding the university.

Naturally, these sites are not impermeable, and many forms of learning will cut across them: for example, a dissertation project may involve data collection in the community, but also classes in research methods or a work placement involving subsequent reflective writing for assessment.

In conjunction with these three sites of learning, we can identify several dimensions of variation in how climate change appears in the curriculum. In order to understand more fully the nature of the spaces – the kinds of learning possible, the distribution of those opportunities, their relationship to other learning opportunities – we need to look beyond the surface features to their underpinning characteristics. The 12 curricular 'features', therefore, highlight structural characteristics of these learning spaces, corresponding to structural features of physical topography (such as elevation, relief or landforms). These features cluster according to three key areas: *access* – who is able to engage in the curriculum, and the conditions in which they do so; *ownership* – the locus of influence, control and decision-making over the curriculum; and *connection* – how the curriculum relates to the field of knowledge and other knowledge areas, to other functions of the university and to practice in wider society. These three areas allow us to assess: on the one hand, the *educational* dimensions of the curriculum – the learning acquired and the meaning it has for learners, the agency that learners express and the impact on their lives; and on the other hand, the *social justice* dimensions – the fair distribution of opportunities for learning, and epistemic recognition in the context of diversity.

The specific curricular features, grouped under access, ownership and connection, are as follows.

## Access

- *Availability:* Is the provision available for all students, or only for those on certain courses? Is it potentially universalisable, or necessarily restricted to limited numbers?
- *Voluntariness:* Is the provision compulsory for all students, compulsory for some students, or voluntary? If not officially obligatory, are there de facto forms of compulsion?
- *Continuity:* Is the learning ongoing throughout students' trajectories at university, available in a specific year, or a 'one-off', appearing at particular moments?

## Ownership

- *Agency:* Who initiated the activity? Is the initiative university-led, lecturer-led or student-led? Who participates in and controls the decision-making process or the content of the activities?
- *Malleability:* How open or closed is the content to modification and development? Is it predefined or constructed during the activity?

- *Certification:* Does the activity lead to the obtaining of credits leading to a degree? Or is there any other formal recognition of the activity that can provide exchange value for students in seeking further study, employment and other opportunities?

### Connection

- *Disciplinarity:* To what extent is the activity linked to a specific academic discipline, and oriented around disciplinary bodies of knowledge, principles of enquiry and methods? Or if not, is it multidisciplinary, interdisciplinary or transdisciplinary?
- *Collaboration:* What level of collective working does the activity involve? Is it a lone process of learning, or is there interaction between student and lecturer, between peers or between multiple forms of actor?
- *Experientiality:* Are there experiential elements, involving learners' participation in real-life situations? Or alternatively is it preparatory learning, or involving simulations?
- *Application:* To what extent is the learning abstract and theoretical, or applied to contexts of practice?
- *Embeddedness:* Is the provision part of an existing taught course or research study or other university programme? Or is it a freestanding activity?
- *Transmodality:* What connections are evident between teaching, research, community engagement and campus sustainability? How porous or isolated are these areas from each other, and to what extent are positive synergies generated?

To comprehend how this frame can be used to understand initiatives in practice, we can take the example of a university running a small grants competition through which students can apply to run a climate action project of their own design. In terms of the spaces, this initiative would be located in the campus or community circles. As to the access features: it is potentially available to all, although only a small number of grants are awarded, so it does not reach all students; it is entirely voluntary; and it is a one-off activity with a fixed timescale. In terms of ownership: while the programme is designed by staff the work is entirely student-led; knowledge is constructed during the activity rather than being predefined; and it carries no course credit. Finally, in terms of connection: it may be disciplinary or multidisciplinary, depending on the proposal in question; it is conducted in collaboration; the learning is experiential;

knowledge is applied rather than abstract; it is independent of formal courses studied; and there is a high level of cross-fertilisation between teaching, research, community engagement and campus sustainability.

The intervention is, therefore, characterised by a high degree of learner agency, with students designing and managing the experience, and has opportunities for interdisciplinary learning, application to real-world problems and possibilities of collaboration – elements that are frequently absent in traditional university courses. Nevertheless, the experience is restricted to a relatively small number of students and is not integrated with the rest of the curriculum, so, in order to provide a rich learning topography for all students, a university would need to be combine this initiative with other forms of experience.

Some other common initiatives are displayed in Table 7.1. These are as follows: a cross-cutting module on climate change, available to all first-year students regardless of the course they are enrolled on; an initiative to revitalise the curriculum of philosophy degrees through embedding contemporary dilemmas of ethics relating to the environment and social justice; an outreach project through which final-year students conduct their dissertations together with local agricultural communities to support them in adapting to changes in climate; and the involvement of students in a consulting capacity in the design of a new zero-carbon student building.

This frame can serve as a tool for mapping existing provision in universities: for determining the location of learning activities, and their characteristics. It draws our attention to a range of crucial questions, such as the connections between the diverse modalities of the university, as explored in Chapters 4 and 5: education, research, services, public debate and campus operations. The most common relationship discussed is that between teaching and research. The 'connected curriculum' framework (Fung 2017) is an important model for higher education generally, in drawing out the benefits of a closer integration between these two. But it has particular relevance to climate change, given the rapid growth in knowledge in the field, its contested nature, the need for experiential learning and for bringing impact to the broader society, all of which make engagement through enquiry essential for students. Facer (2020) also highlights the possibilities of climate change as a 'shared enquiry' between students and educators, moving beyond either student-led or lecturer-led curricula, and breaking down the boundaries between teaching and research. The curriculum topography approach can help us assess and plan for these broader learning opportunities beyond conventional taught courses.

**Table 7.1**: Topographical features in four curricular interventions.

| | Cross-cutting module | Philosophy curriculum | Outreach with agricultural communities | Design of new student centre |
|---|---|---|---|---|
| **Spaces** | | | | |
| **Classroom** | / | / | | |
| **Campus** | | | | / |
| **Community** | | | / | |
| | | | | |
| **Features** | | | | |
| **ACCESS** | | | | |
| **Availability** | Available to all | Only philosophy students | Only certain courses | Available to all |
| **Voluntariness** | Voluntary | Compulsory | Compulsory | Voluntary |
| **Continuity** | First year | Continuous | Final year | One-off |
| **OWNERSHIP** | | | | |
| **Agency** | Staff-initiated and led | Staff-initiated and led | Staff-initiated, student-led | Staff-initiated, student-led |
| **Certification** | Credit-bearing | Credit-bearing | Credit-bearing | No certification |
| **Malleability** | Content predefined | Content predefined | Content constructed | Content constructed |

| CONNECTION | | | | |
|---|---|---|---|---|
| **Disciplinarity** | Non-disciplinary[1] | Disciplinary | Disciplinary | Non-disciplinary |
| **Embeddedness** | Independent of degree courses | Part of degree courses | Part of degree courses | Independent of degree courses |
| **Application** | Abstract and applied | Abstract | Applied | Applied |
| **Experientiality** | Preparatory | Preparatory | Experiential | Experiential |
| **Collaboration** | Individual with some group work | Individual with some group work | Individual and collaborative | Collaborative |
| **Transmodality** | Teaching/research | Teaching/research | Teaching/research/ community engagement | Teaching/campus operations |

[1] The term non-disciplinary is used here to indicate either multidisciplinary, interdisciplinary or transdisciplinary.

The involvement of students in community outreach and university infrastructure is also crucial, both from the perspective of their learning and in enhancing the work itself. Reimers (2021) highlights another important linkage through the role of university students in promoting climate change education beyond the university, particularly in schools. He argues that – given the bottleneck of teacher capacity in delivering effective climate change education, and the limited abilities of ministries of education to resolve the issue – universities and their students have a vital role in working with the lower levels of the education system.

Another important question is that of how climate change links in with other relevant curricular areas – education for sustainable development, disaster education and so forth. These need to be thought of together, not only from the perspective of curriculum congestion, but also so as to bring out the crossovers and synergies between them. A mapping activity such as that encouraged by the curriculum topography approach should also assess how to make the most of these interlinkages, through the diverse configurations of space, form and discipline. For addressing climate change, it is particularly important to bring out these intersections – for example, between the environment and gender equality, racial justice, migration and conflict.

There are no fixed normative implications from the location of an activity on the curriculum map. For example, there is no predetermined hierarchy between classroom, campus and community, or between embedded or freestanding activities, or theoretical and applied learning. In many cases, a diverse spread will be the most appropriate, allowing for different forms of learning to take place and for different types of learner to identify activities with which they can meaningfully engage. There will be inevitable differences between institutions in relation to which kinds of activity they can and should pursue. Yet a marked clustering of activities may be a cause for concern: for example, if none of the climate-related activities in the university are connected with the formal credit-bearing taught provision, or if there are no opportunities for students to develop associational student-led activities outside the classroom.

Having said this, each of these features does have implications for the outcomes of the educational initiative being mapped. As outlined in the framework of curricular transposition (McCowan 2009), slippage of an educational initiative can take place at three stages: in its design, in its implementation in practice, and in its effects on students. Decisions taken across the above features will have a significant impact on these three stages. Designing a climate change intervention from the perspective of chemistry, but without including other disciplinary perspectives (geography, economics, sociology and so on), would have clear implications for the content and impact of the work. Creating a credit-bearing module that cuts across all disciplines would ensure implementation in practice but might lead to resistance from lecturers who had to reduce their core course content to make way for it. Compelling students to attend a course, and providing reward in the form of credit, would inevitably influence the way they related to the content, and potentially enhance or undermine, or in any event change, the way they absorbed the material.

The scheme above has assumed a 'traditional' campus university, with face-to-face teaching, postgraduate study and research, although it can be applied to any kind of higher education institution. Major growth has been seen in recent years in teaching-only institutions and online providers, which have markedly different models of operation but can nevertheless incorporate learning about climate change in their curricula in creative ways. Nevertheless, it needs to be recognised that virtual institutions, or those without campuses or research and community engagement activities, face constraints in their ability to offer the broad range of learning opportunities beyond formal taught courses.

The curriculum topography approach focuses on learning located within or instigated by the university. But it cannot be a closed system, since forms of learning for students are of course not confined to the university. Students are simultaneously learning through their interactions with friends and family, through media and social media. Youth movements, alternative education providers and community associations all provide opportunities for experiential learning outside the university (Facer 2020). There are also increasing opportunities for structured learning from other providers, with MOOCs on climate change now being provided on an open-access basis via platforms such as EdX, Coursera and FutureLearn, from institutions in various countries, such as the University of Helsinki, Wageningen University and the University of Exeter.

No account of the curriculum is complete without some mention of the hidden elements: those aspects of the organisation of the educational institution and its activities that – while unintentional and possibly unbeknown to the architects – nevertheless have a significant influence on learners and their learning. In the case of an institution that endorses climate action, it is essential that there is not a disjuncture between the espoused messages of environmentalism and social justice, and the workings of the institution in practice – for example, investments in fossil fuel companies or not providing a living wage for employees. These contradictions are often evident to students and create an atmosphere of cynicism. At the level of the classroom, and in a more positive vein, lecturers can support the promotion of climate action through the *embodiment* of values, and the creation of a conducive environment of mutual respect, inclusion and commitment to equality. Having said that, it is dangerous to assume that students will necessarily absorb the underlying culture and hidden messages: in some cases, the lack of action in an institution, or its contradictory practices, may in fact inspire students to take action and enhance their learning. Questions of hidden curriculum will be discussed further in Chapter 8.

## Implications for university practice

The primary argument put forward in this chapter is that we should not think about climate change provision at the university primarily in terms of 'teaching' it. This is not because there is anything wrong with teaching, or that teaching should not be included – on the contrary, it is an integral part, as outlined in the previous chapter. But the responsibilities of universities in relation to climate change are those of providing a conducive learning environment. This environment will involve spaces of teaching (experts and facilitators guiding students through the topic), but also of peer learning (students engaging with each other) and self-directed learning (students exploring the ideas on their own) and experiential learning (through actual participation in climate action). It is a shared venture, as Facer (2020: 46) states:

> This is not, then, a question of teaching 'about' sustainability. Instead, it is about creating educational spaces in all programmes in which lecturers and students can work through their field of study to inquire together into the broader questions of what human agency and responsibility means in these conditions.

There has been a welcome increase in climate change courses in universities, and many subject areas are now integrating climate issues into their syllabi and professional requirements (see Nugent 2021). These changes are essential but are just part of what is needed for a learning topography in higher education. Attention is needed to the diverse forms of learning necessary in relation to climate change – ones that can be conceptualised as learning *about*, *for* and *through*. A balance is needed between knowledge of science and debates on climate change, skills relating to mitigation and adaptation and political processes associated with them, and experiential learning of real forms of engagement on campus and beyond.

The formal taught component will best comprise a combination of, on the one hand, discrete dedicated provision (namely, a module specifically on climate change), which would be available to all students on a voluntary basis, and, on the other hand, embedding of material related to climate change across all disciplinary areas, as appropriate for each subject. Given the instrumentalised nature of higher education, and the life implications of university diplomas, making some activities credit-bearing will be expedient in terms of uptake – although there should also be a range of non-accredited activities. Universities should make available organised extracurricular activities for students to involve themselves in environmental issues, but also allow space for students to develop their own actions, unmediated by the institution. Outside the university, opportunities should be provided for students to engage in voluntary work, internships and other work experience relating to climate change. The curriculum topography framework outlined above provides an analytical lens through which we can understand this distribution of activities.

As outlined in the previous sections, there are broader curricular principles relevant to the construction of this learning environment that must be borne in mind. While there are a variety of different legitimate and effective ways in which climate change can appear in the curriculum, the following four criteria are fundamental:

1. *Criticality:* For education to be education, rather than training, conditioning or indoctrination, it must have a critical element. At base, this means that students are enhancing their capacity to make autonomous choices about the validity of claims to truth and value, rather than absorbing in an unquestioning way a predefined bundle of content. Students apply their critical sense during the learning process, and also enhance their capacity for criticality outside of the learning space.
2. *Non-coercion:* The provision must be non-coercive in the sense that it is not imposed on students in either a de jure or de facto way. There are principled and pragmatic reasons for making learning about climate change optional for students. There are constraints on the university's ability to present a unitary set of values, and compulsion in learning is rare at this level. In practice, required modules very often foster resentment in students, and could potentially lead to

a weakening rather than a strengthening of students' commitment in this area. Having said this, in the context of accredited taught programmes, there will inevitably be some compulsion to attend certain courses or display certain learning in assessment; this should be set in the context of students having freely chosen the overall course of study, and a non-coercive environment within the course.

3. *Students' existing knowledge:* Any curricular provision must build on students' existing knowledge and experience of climate change. Children and young people have been highly active in campaigning and mobilising around this theme – as shown by the Fridays for Future and Extinction Rebellion mobilisations – and many university students will already have knowledge, skill and experience in this area. Failing to build on existing knowledge would be poor pedagogy in any event, and is again liable to alienate more than engage students.
4. *Epistemic pluralism:* Higher education has manifested significant homogenisation globally, and needs to remain open to the variety of knowledge traditions, languages and worldviews, as well as to diversity of forms of knowing within cultures – as highlighted by recent movements for decolonisation. Mainstream Western academic knowledge has much to offer, but needs to be placed in dialogue with other knowledge traditions.

These normative considerations – which are generic to university education as a whole, but have relevance in relation to climate change education specifically – provide basic principles on which a diversity of provision can be built.

While this chapter has to a large extent assumed that lecturers and universities are free in creating teaching content, as emphasised by Fahey (2012), there are a number of pressures from different sources in this regard. Constraints on curriculum will be provided by national governments, quality assurance procedures, professional associations and accreditation requirements, and from tradition itself. Furthermore, if constructive alignment is not ensured, assessment may end up undermining the richness of the broad curriculum, with students focusing only on the activities that will gain them points. In addition, there are obvious constraints on the resources available to universities, and to students in pursuing different opportunities. The creation of a curriculum topography of the kind outlined here will, therefore, not be straightforward and require significant commitment from institutions and staff.

Literature on climate change education in schools (for example, Reimers 2021; Rousell and Cutter-Mackenzie-Knowles 2020) argues that teacher education is a key constraint, since this is the mechanism by which official curriculum pronouncements are translated into effective delivery in the classroom in practice. In higher education this is no doubt the case as well, though with an added complexity, in that lecturers are generally assumed to have autonomy over what they teach (with the exception of some cases such as for-profit chains) and rarely have adequate support for their teaching practice. Yet a curriculum topography model releases us from the need to create a 'teacher-proof' curriculum and deliver climate change education *through* lecturers: with a wide range of sources of learning for students, it ceases to be problematic that there is variation in the treatment of climate change in the formal curriculum. Climate change education is delivered through taught courses, but also through a range of other spaces, both formal and non-formal in the university campus and beyond.

This environment does not then need to be perfectly aligned, with everybody 'singing from the same hymn sheet', as is often prescribed. In a university with academic freedom, there will inevitably be different perspectives on a topic like climate change among lecturers and students, even if they all believe in its existence, take it seriously and are seeking solutions. Far from being an impediment, the unevenness or even messiness of messaging is part of the rich learning environment, in introducing students to those diverse perspectives and allowing them to build the autonomy to frame their own beliefs and commitments in relation to them. The curriculum topography for climate change is more like an organic landscape garden than a formal symmetrical one.

Haydon (1977) makes the important point that the right to education is not only one of provision, but also of non-interference: states need to ensure not only that they present opportunities for learning through educational institutions, but also that they do not prevent their populations from educating themselves, through censorship, or otherwise restricting the time or resources people have for accessing information and learning opportunities. We can apply similar ideas to the university in relation to climate change. One of the duties of HEIs is not to get in the way – either by crowding out available time with the formal part of the curriculum or in other ways obstructing or devaluing extracurricular learning. Students will naturally create opportunities for these forms of sharing and learning if the conditions are right. These may be formed on a physical campus, but if a physical campus is not possible, then alternative kinds of space, including virtual, can serve this end.

Furthermore, it must be recognised that – despite the substantial attention to sustainability and climate change in universities in recent years (McCowan et al. 2021) – many HEIs remain indifferent, and some may even oppose these aims. This unfavourable environment does not, however, preclude the establishment of opportunities for learning about climate change, in part due to the diverse and decentralised nature of HEIs and the relative autonomy of lecturers in their teaching and students in organising their own learning. As mentioned above, there may also be opportunities for students to develop important capacities through their *opposition* to their institutions.

The value of the university as a learning experience, therefore, resides as much in the arena that it provides, as in its intentional programmes of study. The meetings across diversity, the cohabitation of philosophies, foci and disciplines, conducted in the spirit of open enquiry and the quest for understanding, provide an unparalleled space for the development of a new human and ecological paradigm. The emergent possibilities of this space for dialogue and exploration, and the arising of unanticipated ideas, connections and becomings, will be vital if we are to retain any chance of addressing the environmental and social crises facing humankind. These ideas are explored further in the chapter that follows, as the focus widens to consider the ways in which the whole institution of the university engages with climate change.

## Notes

1 PISA is the OECD's Programme for International Student Assessment, which tests fifteen-year-olds' skills and knowledge across 81 countries.

2 https://carbonliteracy.com/

3 The three spaces of classroom, campus and community were discussed in a rudimentary form in McCowan (2014).

# 8
# Institutional embodiment

'But where is the university?', asks the fictional visitor to Oxford in Gilbert Ryle's often quoted story in *The Concept of Mind* (1949). Having seen the libraries, departments, lecture halls, student accommodation and playing fields, he is unsure where the institution itself is. Like the parable of the blind men and the elephant, Ryle's vignette is pointing us to the category mistake of trying to find the whole in the parts, and there is something particularly complex about the university in this regard. As organisations go, it is unusually disparate in its members, functions and mission. Those working within universities have high degrees of autonomy, often maintaining their strongest allegiances for ideas and bodies outside their own institution (for example, the community of historians of the medieval period or adherents to poststructuralism), and the largest constituency (students) spend only a few years in the institution before moving on. Mission statements try to bring some unity of purpose to this kaleidoscope, but are necessarily vague and rarely unite in practice.

Does this creative disorder militate against aligning the institution with sustainability and climate action? Are leadership and institutional strategy powerless in the face of this diversity of interests and autonomy of action? Certainly, we have to be realistic in our expectations. Especially in larger public or philanthropic institutions (commercially oriented and for-profit universities are exceptions in this regard), aligning all the functions and activities of the organisational subcomponents and actors with a common goal is highly difficult, and even if it were possible, students and academics would still be thinking differently – and therefore produce unaligned outcomes of teaching and research. From a normative perspective, we also need to guard against command and control. A military-style university in which unquestioning allegiance and obedience were achieved would in fact be a very inadequate institution. There is something about the need for agency, imagination and creativity in all processes of learning and research that makes commanding their achievement from the outside unthinkable.

That is not to say that leadership is unimportant. The structures and cultures of our institutions still have a strong influence on what happens within, and leadership shapes them in important ways. An often cited example is that of Michael Crow at Arizona State University, who brought about important shifts towards a sustainability-focused institution by establishing interdisciplinary centres for both research and teaching, and promoting purposeful and impactful scholarship (Crow 2010; Newman 2021). At the same time, care is needed with the 'great man' approach to history: many invisible people also contribute to these profound transformations, and sometimes the figurehead may be presiding over monumental changes, rather than instigating them.

Whatever the dynamics involved, attention to the institutional is crucial. This chapter will argue that we should approach institutions through the notion of 'embodiment'. We can think of embodiment at the individual level – how people incorporate in their everyday lives and lived experiences the principles and commitments they adhere to – but also in terms of the body of the institution. Universities should incorporate their commitments and actions on climate within themselves, in their management, their physical estates, their modus operandi and their relationships. Similar ideas are expressed by the notion of 'embeddedness': referring to 'sustainable organizations that have undergone a paradigmatic shift and adopted firm-wide sustainability embeddedness so that sustainability has become an organizational way of life' (Le Roux and Pretorius 2016).

This idea corresponds to the *expressive* rather than the *projective* mode: that institutions should think not only of creating sustainability outside of themselves (in time and place), but of expressing it within their day-to-day activities (McCowan 2023). As climate change is rooted not only in our technologies and modes of production, but also in our social structures and relationships (including our relationship to self), so this embodiment relates not only to the greenhouse gas emissions of universities but also to the cultures that support them.

Crucially, embodiment does not mean homogenisation, standardisation or unification. A university may embody transformative climate action, but still allow space for diverse ways of realising it, and even for those working or studying there to ignore and even resist it. Even if it were practically possible, unifying all the values adhered to and expressed by all members of a university would hardly be legitimate. An unavoidable tension is created therefore by the desire for purpose, direction and impact on the part of the institution, and the diversity, freedom and agency of those working inside – a tension that is important to face up to but, as argued by this chapter, is not necessarily fatal in terms of the university's role in climate action.

Ideas of the 'prefigurative' can be of use here (Boggs 1978; Epstein 1988). Prefiguration is the adoption of principles of the desired end state within the means of getting there, an instantiation of the goal to be achieved. A common example is the use of democratic methods for achieving a democratic polity – or alternatively avoiding non-democratic, coercive or violent means of achieving it. An earlier study on school-level education (McCowan 2010) identified three roles that a prefigurative approach within education can play: as an inherent good, as an exemplar and as a learning experience. To these, we can add a fourth in this case, as a site for experimentation. Applied to the climate crisis, these four roles would be as follows: (i) conducting effective climate action in the here and now; (ii) serving as a positive example of climate action of inspiration to others; (iii) providing an experience of rich learning about climate; (iv) acting as a space for innovation and experimentation on sustainable living. These four aspects will be outlined in the sections that follow.

## Prefiguring the climate transformation

### Climate action in the here and now

Universities are communities in their own right. They are somewhat different from conventional geographic communities since their residents are mostly temporary, with students staying only for a few years, and staff only part of their week – in most cases, also occupying other geographical communities. Nevertheless, with these special characteristics, they do represent places with their own distinctive interactions and impacts, including positive and negative influences on the environment.

The most obvious significance of embodiment is that it is, in itself, a good. Embodying or prefiguring climate action in an educational institution is actually 'doing' climate action in that particular space. In the case of universities, this is not a negligible space at all: as the Shields (2019) study shows, the carbon emissions from international student flights globally are equivalent to a country the size of Tunisia or Croatia. The total greenhouse gas emissions are very significant. Emissions from higher education in the UK were estimated at 3.3 million tonnes in 2005, a rise of 33 per cent since 1990. A total of 2.3 million tonnes of carbon in 2012–13 was omitted from energy consumption alone (Mazhar et al.

2017). A more recent comprehensive report (Royal Anniversary Trust 2023) has put the total emissions of the sector at 18 million tonnes for the 2021–2 year, 2.3 per cent of the entire national emissions of the UK. Of these, the largest proportions are 36 per cent for supply chain, 19 per cent student accommodation, 12 per cent student flights, 8 per cent student commuting and 5 per cent construction.

HEIs around the world have been active in trying to reduce their emissions of greenhouse gases. The first step in this process is to know what those emissions are – a task which, despite being the most direct and concrete of the climate-related impacts of the university, is nevertheless challenging. Institutions have generally used their own forms of measurement, borrowing from the array of tools available in the private sector, although proposals have been made for unifying and providing a common gauge for all (Helmers et al. 2021; Ozawa-Meida et al. 2013; Robinson et al. 2018; Redfern and Zhong 2017). In addition to the challenge of finding tools that are comparable across institutions, another constant barrier is the fear of negative publicity for institutions, in a highly competitive market environment.

Institutional emissions are generally divided into the categories of Scope 1, 2 and 3:

> Scope 1 [refers to] direct carbon emissions that occur from sources owned or controlled by the organization and scope 2 accounts for emissions from the generation of purchased electricity. Scope 3 is all other indirect emissions that arise as a consequence of various organizational activities, but occur from sources not owned or controlled by the organization. (Mazhar et al. 2017)

Scope 1 and Scope 2 emissions are easier to calculate. The former involves assessing the direct fuel usage in estates and institutional vehicles. As regards purchased electricity (Scope 2), providers can now give information on the quantity and also the mix of fuel sources used. In England and Wales, it is now compulsory for HEIs to provide varied information on Scope 1 and Scope 2 emissions to the Higher Education Statistics Agency (HESA n.d.). Scope 3 is much more difficult as institutions are reliant on obtaining the relevant information from the companies from which they are sourcing their products – a considerable task given the huge quantity and variety of products in question. Nevertheless, Scope 3 are the most substantial, with the Royal Anniversary Trust (2023) report estimating 88 per cent of all higher education emissions coming from this source, compared with 7 per cent from Scope 1 and 5 per cent from Scope 2. A further area that could be included in carbon accounting from HEIs is 'financed emissions' – pension schemes, endowments and investments (Royal Anniversary Trust 2023).

In understanding the impact of higher education on emissions, it is important to disentangle which of these involve substitution and which extra impacts. If a student is living at university instead of at home, then the general living emissions may be the same, but they are just transferred from one place to another. However, the construction of research labs and flights from student mobility involve extra emissions that would not otherwise have been incurred.

Once emissions have been calculated, what if anything can institutions do about them? Concrete actions can be taken across all three of these areas. In relation to Scope 1, reductions can be made in fuel usage: in transport, heating and cooling of estates. Emissions can also be reduced through conversion to electric vehicles, and encouraging staff to walk and cycle. Some significant strides have been made in relation to the design and efficiency of buildings, and therefore to reduction of energy usage. My own institution, UCL, for example, built a new student centre in 2019 which was awarded BREEAM 'Outstanding' for its environmental performance, using lower-carbon building materials and natural light, heating and cooling to reduce external energy usage. There are many other buildings of this type being constructed in universities around the world. However, the real challenge is with existing buildings – particularly when these are many decades or centuries old. Retrofitting buildings can be very expensive, and on occasions limited by architectural and heritage considerations. Extreme weather may increase energy usage – for example, with longer heatwaves or new areas experiencing prolonged heat (or sometimes cold) – leading to a vicious cycle of the causes and impacts of climate change, as in so many other areas.

As regards Scope 2 emissions, to some extent institutions' hands are tied by the mix of sources of electricity available in the locality, although in some cases it is possible to request from one's energy provider either renewable only energy sources, or a higher proportion of renewables. This choice may come with a 'green premium' (Gates 2021), so it is not one that would be welcomed by all institutions.

Finally, there is Scope 3, the hardest of the three both to measure and influence. The quantities and type of products consumed by institutions, and other upstream and downstream impacts, form the vast majority of an institution's carbon footprint and are therefore crucial in reducing their emissions. Procurement policies can reduce carbon footprint, not only in terms of reducing overall consumption, but also in choosing different products and between producers of the same product on the basis of their environmental impact. One example that is commonly seen is the use of plant-based products in catering for events in universities, given the significant carbon footprint of livestock, particularly beef. These changes are difficult if the green premium is high, given that universities are often operating in a tight financial environment.

International travel has already been discussed in Chapter 5. Efforts have been made to reduce staff travel for research meetings and conferences, aided significantly by the culture shift and developments in technology that took place as a result of the Covid pandemic. In certain disciplinary areas there have been concerted movements to change attitudes and practices (for example, Gill 2021). Videoconferencing is now the norm and has led to some significant changes: in Brazil, for example, doctoral examining used to involve significant domestic air travel, but has now been almost completely replaced by virtual *vivas*. Nevertheless, few people are convinced that virtual interactions are equivalent to in-person ones when it comes to building relationships and trust, meeting people and networking, and for immersion in different cultural contexts, so travel for research and conferences will continue, even if to a reduced degree.

International student travel, on the other hand, has returned to something like business as usual after the disruption seen during the Covid pandemic. Barriers to change in this regard are not only the vital importance of the income for HEIs in market systems (such as USA, UK and Australia), but also the ambiguity of the shared responsibility. Are the emissions part of the carbon footprint of the institution or the student, and on whom does the onus of change fall? In a marketised system, in which the survival and thriving of institutions may depend on the recruitment of international students, can we realistically expect them to cut back?

As argued throughout this book, the climate crisis is one rooted in human communities and the human being, so the transformations needed to address the crisis will also bring about changes in these spheres. These changes also have intrinsic worth. Movements away from exploitative relations based on self-interest are good for the climate, but are also positive in terms of creating peaceful, harmonious and enriching communities in which all people can live in dignity. University initiatives that are relevant in this regard include equity, diversity and inclusion (EDI), and antiharassment and bullying policies, creating a level playing field and an environment for mutual enrichment for all (students and staff) in relation to gender, race/ethnicity, abilities, social class, religion and sexuality. Despite having removed most formal bars (such as the ban on Catholics that existed in England until the mid-nineteenth century), universities have a long way to go in this regard, with de facto barriers to both entering and succeeding. Kinol et al. (2023), for example, argue that universities can apply the Green New Deal policy framework to enable climate justice within the institution, including non-extractive hiring, operationalising antiracist principles and resisting financialisation.

In the short term, addressing the climate crisis unfortunately is more about stopping doing bad things than starting good things – there is no getting away from the fact that the primary task is not to emit greenhouse gases. As discussed in Chapter 2, the only other way forward is to either remove carbon from the atmosphere, or to make other adjustments that will counteract the negative impacts (for example, geo-engineering). At present, neither of these alternatives are sufficiently developed or safe to provide a viable way forward on their own. Nevertheless, carbon capture must certainly be part of the solution, in particular through natural means (plants and trees) given the other environmental benefits it brings. Sustainable management of natural areas, tree planting and protection of existing forests is therefore an important role of universities, both on land they own and in providing these services elsewhere. These efforts form part of the broader task of regeneration, involving both human and non-human communities, through which both climate change and other environmental challenges such as biodiversity loss can be addressed and the conditions created for a liveable planet.

## Serving as an exemplar

The above section has covered the direct impacts that universities bring, corresponding to the movement directly from left to right of the diagram in Figure 4.2, straight from university to ecosphere. Yet there are other roles that embodiment plays. One of these is to serve as an example – and possibly a beacon of hope – for other people, organisations and institutions.

Concerted climate action in society is extremely hard to achieve, since the immediate costs appear high for those in privileged positions: for individuals, reducing consumption and changing habits in ways that may appear inconvenient, and for organisations, making structural changes and possibly reducing profits. Making the shift, therefore, requires strong support from all segments of society and the changing of habits in ways that might initially be irksome, but will ultimately become natural (such as putting on a seatbelt in a car or bringing one's own bags to the supermarket). For universities to disregard the natural environment and the dangers of climate change undermines these collective efforts, and gives other organisations an excuse to take their foot off the pedal. Small things can be important. Most of us have had the experience of working in institutions with lights blazing and computers whirring all night in deserted offices, or even of heating and air conditioning battling against each other!

Hypocrisy is particularly insidious (Orr 1994). An institution that promotes environmentally friendly messages and trumpets its green credentials while simultaneously exploiting and polluting is particularly dangerous as it promotes cynicism among the general population, leading to a generalised sense that no powerful organisations are sincere in their efforts, and undermining the work of those organisations that are. The implications of these dynamics for learning will be addressed in the section that follows, but they are also relevant for those observing the institution from the outside.

Fortunately, universities also function as positive exemplars in this regard. As stated in the previous section, there are many inspiring examples of transformations of campuses, construction of sustainable buildings, campaigns and mobilisations, and many other actions. Fossil fuel divestment is an important instance here. A few universities globally have substantial endowments, partly invested in stocks and shares (it is important to emphasise that it is a small minority of HEIs, with most either being owned by the state or privately owned but without significant endowments). Fossil fuel companies have been part of the portfolio of these investments, as they are for many funds. Starting in the early 2010s there have been concerted campaigns – particularly from students – to pressure institutions into removing companies associated with fossil fuels from these portfolios, and they have had significant success (Stephens et al. 2018; Treisman 2021). Of course, there is some intrinsic good here (if it is indeed the case that divestment results in a reduction of exploitation of fossil fuels), but very often these campaigns serve the primary purpose of raising awareness of the damage caused by these companies and promoting broader disinvestment across other organisations (Grady-Benson and Sarathi 2016).

Beyond the direct impact of these campaigns and mobilisations, there is also an exemplar effect in building movements, solidarity and coalitions, and showing the possibilities of collective action in relation to climate. While some may dismiss them as rhetorical, we can also see the importance in this regard of declarations of the climate emergency. A number of institutions in the UK, starting with the University of Bristol in

2019, have made public announcements of the seriousness of the crisis and their commitment to addressing it. According to Latter and Capstick (2021), these declarations have a largely promotional role, without a clear connection with concrete action, although together with the large numbers of local governments that have also declared emergencies, they have certainly led to greater visibility of the crisis among the general public.

## Providing a rich learning experience

Spaces of embodiment also provide distinctive learning experiences. Learning is a result not only of designed and structured activities (formal and non-formal education) but also of experiences and interactions in day-to-day life (informal learning). While these environments may be a source of knowledge and skills, they are particularly important for shaping values. What we think of as culture is the broadest and most powerful of these learning environments, affecting all of us.

In many cases, the nature of the environments we move in may be apparently random, unintentional or just part of the general culture. But there are cases in which spaces or institutions can provide specific environments that – while usually invisible or unstated – are either intentional or, if unwitting, at least inherent to the purposes of that space. In educational circles, this phenomenon is known as the 'hidden curriculum', as opposed to the official curriculum (the one contained in policy documents and syllabi) and the unofficial or taught curriculum (the one actually implemented in schools). There is a fair body of literature on the hidden curriculum going back decades, with both functionalist and Marxist strands, assessing how educational institutions provide a deep formation of young people for their roles in society, promoting variously conformity, competition, acceptance of hierarchy or entitlement (Giroux and Penna 1979; Hinchcliffe 2020; Jackson 1968; Margolis 2001).

Ideas of the hidden curriculum in HEIs have clear application to climate change and sustainability (Winter and Cotton 2012). Regardless of what is formally taught in classrooms, institutions provide environments that display different kinds of practice, value and relationships that influence those inside them. For example, institutions can model good practice in responsible consumer choices through their procurement policy, making hard choices about avoiding the cheapest products when they are socially or environmentally harmful and avoiding waste and energy inefficiency, in addition to actions focused on human communities such as paying employees a living wage, ensuring equity, diversity and inclusion, and so forth. These visible actions and policies are underpinned by more subtle expressions and cultures of competitiveness, exclusivity and intolerance, or alternatively cooperativism, inclusion and open-mindedness. It is not infrequent that external trappings of greenness are accompanied by contradictory underpinning currents, particularly when being green can provide market advantage in bringing positive brand associations and student recruitment.

As argued by Cotton et al. (2020: 30), 'there is increasing evidence that much sustainability education occurs "under the radar" through informal processes such as the hidden curriculum and place-based learning'. She identifies three spheres in which the hidden curriculum can be identified:

> 1. The formal curriculum/syllabus: choices about what to include and exclude from the syllabus
> 2. Pedagogy/teaching approaches: sending messages about a tutor's underlying values
> 3. Institution/campus environment: hidden curriculum of place. (p.31)

In relation to the first of these points, the limited presence of sustainability in formal curricula displays and communicates to students the low value placed on it (although the commitment, curiosity and engagement of those who do incorporate sustainability in their teaching may be profoundly inspiring). It is the third of these points, however – the institution and campus environment – that receives the most attention. For example, marketisation – value for money and institutional competition – forms part of the deep hidden curriculum backdrop of HEIs in the UK.

Contrary to many commentators, Cotton et al. (2020) see the hidden curriculum as potentially positive, despite its sinister label. In fact, positive hidden curricula are commonplace in child-rearing practices, where carers attempt to model good behaviour, and create an environment in which kindness, sharing and politeness are part of the fabric of the everyday. The campus and institution can be used to foster a conducive environment, as shown in the '7 Steps to Taking an Institutional Approach to Learning about Sustainability' of Plymouth University (2015), using the architecture, processes and strategies of the institution to generate learning. Students are sometimes unaware of their universities' sustainability actions, so institutions could do more to publicise them – for example, through simple signage.

The aim here is not to promote an idea that all initiatives be led by the central university organisation; a positive hidden curriculum must also instil a sense that action can and must be taken by all. In part, this task involves concrete elements of space and opportunity, but also the development of a deeper sense of self – as illustrated in Paulo Freire's (1970) primary pedagogical aim of learners becoming subjects, rather than objects of processes outside of their control. As discussed in Chapter 7, an effective curriculum topography can instil a sense that climate action is within the control of students and can be initiated by them. The learning environments of HEIs, therefore, bring together a mix of positive and negative influences, ones that, while elusive, can be shaped by those inside them.

The general assumption of the literature about hidden curricula, and about learning environments generally, is one of correspondence: that a democratic environment will lead to learners becoming democratic, a competitive environment to them becoming competitive and so forth. To some extent this relationship holds, but significant caution is needed in this regard. In some cases, influence will be brought to bear on an individual, but may not be strong enough to bring a noticeable shift (possibly because of the more powerful influences of hidden curricula/ learning environments in other spaces of their lives, such as family, friendship groups, subculture and so on). In other cases, it may actually bring a contrary influence. As explored in McCowan (2009) in relation to citizenship, this outcome is particularly likely in cases in which the school represents an external, imposed authority running counter to the community or culture that the student identifies with. Even in the absence of such conflict in the relationship, there may be scepticism about official messages. This was certainly the case with my own children who were permanently disdainful of the messages of cooperation and sharing they were fed in primary school assemblies, accompanied by daily renditions of uplifting Bob Marley songs! They did actually value cooperation and sharing; they just did not want their school to force-feed them with it.

Can a learning environment be more or less generative independently of being good or bad? In other words, might a less environmentally friendly university spark critical learning opportunities for students more effectively than one in which there were more environmental practices? It is certainly possible, since there is much more to education than modelling: posing questions, challenging assumptions, incentivising curiosity, providing resources for deeper exploration and so forth. A learning environment for sustainability, therefore, needs to encompass both creating a positive hidden curriculum and other generative learning opportunities.

There is much we do not know about the influence of hidden curricula, and for obvious reasons the 'hidden' nature of these phenomena makes them hard to research and complicates attribution. In the absence of a firm empirical evidence base, we must approach the issue with attention and reflection, taking the issue seriously but remaining open to evolving realities. While the impacts are indeed unpredictable, the most coherent approach is still to model the reality that is desired – in this case, a sustainable, fossil-free environment – while taking care not to assume that it will automatically make all those within it into sustainable people. While not being straightforward, deterministic and linear, the hidden curriculum matters and there are strong reasons for supporting embodiment from a learning perspective.

## Acting as a space for experimentation

A final opportunity provided by embodiment or prefiguration is that of experimentation. We do not know exactly how to live in a sustainable world. For sure, we have some examples of sustainable communities, such as those of Indigenous peoples that have lived for long periods of time in harmony and mutual enrichment with the non-human environment. But most contemporary societies will struggle to replicate those forms, embedded as they are in urban spaces with capitalistic market relations and supply chains, and having lost the knowledge needed to thrive in direct contact with nature. Those relations – for most people on Earth – will need then to be recreated.

Experimentation with new forms of living is, therefore, a key part of regeneration and climate action. In part, 'experimentation' is meant in the more technical sense of experimenting with interventions in the climate system (such as use of renewable energies, carbon capture, geo-engineering), trying out new computer models for understanding future climate scenarios and adapting technologies to higher temperature levels. But it can also be used in a more social sense, of experimenting with new configurations of human relationships, and institutionally too, experimenting with new strategies and policies. Finally, experimentation is needed in relation to ideas, opening up the imagination of what communities, societies and educational institutions might look like.

The idea of a 'living lab' has become popular in literature and practice on sustainability in recent years. It refers to a space in which different stakeholders come together to innovate and experiment in relation to real-world sustainability issues (König 2013; Evans and Karvonen 2014):

> Living laboratories provide a space for multiple stakeholders to address local challenges by jointly framing issues and producing new knowledge deemed by all an adequate basis for concerted action. The purpose of living laboratories is not only to allow novel things to be tried that would not be possible in conventional urban settings, but to also carefully monitor their social and physical impacts in order to provide a robust knowledge base for learning. (König and Evans 2013: 1–2)

It might seem strange to apply this idea to the university, given that the institution is a space for real labs – that is, controlled environments in which experiments are conducted in artificial conditions – and for the generation of theoretical knowledge separated from its real-world applications. Yet the university is more than just teaching and research, and the real-life situations of campuses and estates do provide an opportunity for living labs. These spaces provide possibilities of experimentation that contribute both to the learning of those individuals and groups involved, but also to collective learning in terms of research and scholarship. A key aspect of living labs, in fact, is that they are conducive spaces for bringing together activities of teaching, research and community engagement that are very often held apart (Purcell et al. 2019).

In summary, prefiguring is a good in itself, but it also brings instrumental benefits that can make a significant contribution not only to the climate action of the institution in question, but to society more broadly. It may never be possible to have complete embodiment in any individual institution, given the diverse and mildly anarchic nature of higher education. Furthermore, the balance between these four elements will always vary – so some institutions may focus very much on the experimentation side, while others more on the public relations side (exemplar).

This chapter thus far has focused mainly on the campus operations modality, in which these questions of embodiment are particularly relevant. But ideas of embodiment can also be applied to the other four modalities, as has been seen in Chapters 6 and 7 on teaching, learning and the curriculum, and in the chapter that follows, Chapter 9, on public engagement and services. The next section will draw out implications for the research modality. While this book does not aim to summarise the content of climate research, there are a range of implications for how research agendas broadly in universities should be imagined, managed and practised.

## Embodiment in research agendas

Research is a good example of the relatively unruly nature of higher education, since its focus and the way it is conducted is largely out of the control of institutional management. The factors shaping research agendas are partly visible and concrete – the funds and time allocated, structures of research programmes and centres, staffing and so forth – and partly a question of a 'hidden curriculum' of cultural and social norms influencing values, choices and preferences. These factors affect what questions research focuses on, how it is carried out and how it is communicated to others. The last of these questions will be dealt with in Chapter 9, relating to engagement with external communities.

The material factors of funding and staffing research each display varying configurations depending on context. In some cases, research is one of the stipulated activities of academic staff, built into their profile of work, alongside teaching (and to a lesser extent public service). In others, research time is 'bought' through the achievement of external funding on a competitive basis. There are also systems in which the Humboldtian unity of research and teaching is broken and separated between different staff members. Whatever the configuration, it is clear that only a small proportion of those working in HEIs globally actually have the time and qualifications to do formal research. The rise in part-time and precarious appointments in higher education is well documented (Kahn et al. 2024; Read 2023), and Altbach et al. (2009) estimated that half of higher education staff globally may have no more than a bachelor's degree.

Furthermore, there are the direct costs of research. In cases in which substantial resources are needed (for example, in most empirical work, particularly in science and technology), then funds from outside the university are usually required, often from public research funds or from private foundations. Some aspects of climate research are extremely expensive or require specialist equipment (for example, climate modelling) and are consequently restricted to a few well-funded institutions, or to specialist non-university research centres such as NASA and the UK's Met Office. Funding of research inevitably brings us to questions of neoliberalism, privatisation and the compression of the public space, given the centrality of public funding for protecting research in the public interest.

The strongest influences on methods and content of research are disciplinary communities that provide historical weight and contemporary gatekeeping through peer-reviewed journals. Without doubt, the awarding of research funds is also an important influence

– particularly in those areas in which funding is more plentiful (such as STEM, medicine). But even here the role of the government (or other awarding body) is relatively hands off, with the choices of topics and methods and the decisions on awarding again being made largely from within the academic community – as enshrined in the Haldane Principle. Governments and institutions, therefore, can nudge, but rarely direct. In many ways, this is as it should be, since insights, discoveries and breakthroughs cannot be forced and often emerge spontaneously and serendipitously during long processes of exploration and reflection. Furthermore, those leading research funding processes can rarely be experts in the finer detail of all the areas of research. Nevertheless, disciplines have their own conservatism, so incentivising focus on topics of crucial relevance to the planet, as well as working in new ways across disciplines, can have a positive impact (Rapley et al. 2014).

What does it mean to embody principles of climate action and climate justice within research agendas? In the first instance, it involves simply encouraging research. It may seem a 'no-brainer' for universities to support research since it is research excellence that carries the greatest weight in international university rankings, and is strongly associated with historically accrued prestige. Yet what furthers these ends of status competition is the highly visible, elite research, supported by large grants and publishing in top-end journals. Greater commitment from institutions is needed to promote research activity across the whole of the academic community, including among groups and in academic areas in which there may be seen to be lower 'returns'. Furthermore, there is a broader understanding of what constitutes research – a better term for which perhaps is 'scholarship' – that underpins all academic activities, including teaching and community engagement, and which needs to be protected with time and space.

However, more research does not necessarily mean the right kind of research. In some cases, there are clearly pernicious influences on university research, such as that of the Koch family in promoting climate denial (Kinol et al. 2023). In others, there may be more subtle steers away from public benefits, or away from engagement with pressing societal challenges. The problems of the ivory tower and inert, disconnected knowledge have been widely expressed in recent years – perhaps too much, in ways that unduly devalue the importance of theory and curiosity-driven research. Yet it would be hard to deny that the climate emergency warrants new approaches to science. The primacy of the 'new' in science needs to be accompanied by an emphasis on the 'useful' (Crow 2010; Rapley et al. 2014). Lubchenco and Rapley (2020) argue that scientists need to enter a new social contract, moving beyond the binary between basic and applied research, to what Stokes (1997) refers to as use-inspired research, 'pursuing fundamental knowledge to solve problems that are immediately relevant to societal needs' (Lubchenco and Rapley 2020: 3), and moving beyond simply communicating to the outside world, and engaging with communities and policymakers to create solutions.

There is also the question of connections. Institutions need to support fruitful relationships between research and other areas of university activity, most obviously teaching, but also community engagement and campus operations – as proposed in Dilly Fung's (2017) 'connected curriculum'. The benefits of integrating research into teaching and learning are well recognised (and discussed in relation to transmodality in Chapter 7) but there are also positive influences of teaching on research.

There also needs to be balanced support between disciplinary areas. STEM areas in the university have come to the fore in the second half of the twentieth century, at the expense of arts, humanities and social sciences, following the logic of the importance of industrial and technological development for economic growth. In a distinct trend, marketisation of higher education has led to a rise of applied, professional and vocational subjects (business studies, law, engineering, health, sports), at the expense of traditional academic subjects (including natural sciences). While these trends of supply and demand affect taught courses most directly, they also end up affecting research, with the isolation of departmental budgets, and the cutting of supposedly unprofitable areas of philosophy, history, literature and so forth. As argued in the first part of this book, all academic areas are vital to solving the climate conundrum, and yet research support is very heavily weighted in favour of STEM subjects. Not only do different disciplinary areas need to be in balance, but they also need to work together – for example, with climate scientists engaging with social and behavioural sciences to ensure effective communication and motivate action (Rapley et al. 2014).

Another reason for encouraging connections between different disciplinary areas is the centrality of values in the climate conundrum. Leal Filho et al. (2018), for example, argue for the importance of ethics in climate research, applauding the appointment of a climate ethicist to lead Working Group III (on mitigation) in the 2014 IPCC report. *Laudato Si'* (Pope Francis 2015), as might be expected, also shows a strong binding of values and science, cautioning against faith in technological progress without a moral dimension. Ethics, of course, should be a part of any academic discipline, but also provides an opportunity for philosophers to engage directly with other areas.

Higher education systems should also allow space for different kinds of methodology and epistemology. For the most part, universities are places of considerable freedom in terms of method and belief, although resources, incentives and visibility – in the form of research grants and academic journals – exert some restrictions. In relation to climate change specifically, efforts should be made to promote diversity of perspective and research design in exploring the multifaceted phenomenon, and to maximise imagination and creativity in solutions. This point connects strongly with ideas dialogue of knowledges, and engagement with Indigenous knowledge, discussed in Chapter 3. While challenging for any institution, the university should aim to promote epistemic pluralism, ensuring a diversity of worldviews, between and within cultures, and the use of multiple languages. This process is best achieved through co-production between different stakeholders, as will be explored further in Chapter 9.

As argued in a previous book (McCowan 2019), the quest for relevance and immediate impact should not squeeze out open-ended, curiosity-driven research. The developmental university model has brought important shifts in higher education, moving towards engagement with external communities, prioritisation of social justice and environmental concerns, and attention to application of knowledge for real world issues. Yet these laudable aims should not take the place of traditional scholarship. Effective practice cannot exist without theory, and blue skies research is crucial – not only in more speculative areas of scholarship, but even in applied ones, since abstract breakthroughs very often have concrete ramifications in the medium to long term. Many years passed before the discoveries about the heat-trapping properties of carbon dioxide, initiated by Eunice Foote and John Tyndall, had vital practical application to enable understanding of the impacts of burning fossil fuels (Bell 2021).

In addition to the theory/practice question, there is also that of values. As argued in relation to student learning, a unified value set in research is both an impossibility for and anathema to the university. The large array of people involved in research makes some difference of position inevitable, and academic freedom protects that diversity as a matter of principle. That said, there is no requirement for value neutrality on the part of institutions or researchers; Freire (1970) argues moreover that 'sitting on the fence' is in fact supporting domestication rather than liberation. Universities can vigorously pursue solutions to climate change through research, while remaining open to diverse positions on the way forward.

Promoting high-quality, impactful research, then, is rather like cultivating plants. The magic itself is out of the hands of the institution, in the self-directed inspiration of scholars to enquire, explore and understand. But, just as a gardener would by preparing the soil, and watering and tending to the plants, institutions can create the right conditions, ensuring that all staff have the opportunity to pursue their lines of research in an environment of diversity of perspectives and disciplines, and are able to make fruitful connections with other institutional activities.

This section has focused primarily on the institutional factors affecting research agendas; to these must be added the global trends discussed in Chapters 3 and 5. There are significant inequalities, disparities and skews in the questions of who conducts research, whose voices are heard and what topics are researched, depending on language, historically accrued university prestige and funding. The systematic review on university responses to the climate crisis conducted by Climate-U (Nussey et al. 2023) revealed both the lack of meta-reflections on research agendas, and the barriers that exist to ensuring equity, dialogue of knowledges and engagement across all disciplinary areas (Czerniewicz et al. 2017; García del Amo 2020; Leal Filho et al. 2018).

## Catalysing embodiment

Cultures and practices can arise spontaneously or organically from the bottom up in a human community, but in an organisation that has existing structures and hierarchies, it would be negligent not to try to actively nurture them. Questions of embodiment, therefore, necessarily give rise to those of leadership and governance. At the same time, it is important to acknowledge the limitations of these steering devices, particularly in an institution like the university.

Some aspects of work are directly within the purview of management – for example, procurement policy, conditions of employment, design of new buildings, carbon management planning – but many are more akin to the work of the gardener outlined in the previous section, that is, creating the conditions for flourishing, and then letting the plants and trees grow by themselves. A previous study of sustainability in HEIs across seven countries identified the following three key tasks of leaders: (i) presenting a vision – putting forward an ambitious and disruptive vision to inspire and challenge others; (ii) fostering alignment and synergy – capitalising on the benefits of connecting diverse areas of work; and (iii) nurturing innovation – providing incentives and removing barriers to bottom-up change (McCowan et al. 2021). For Purcell et al. (2019: 1347), the key point is how:

> . . . the senior management hierarchy (SMH; the bounded executive command-control system) interacts with the more agile unbounded community of social networks (CSN; groups that convene around shared purpose projects, including standing committees, task and finish groups and informal assemblies).

These questions are also highly contextual, with quite different conditions for influence between large and small institutions, and between public and private ones. Within each of those sectors there are also important differences between public institutions strongly subject to state control and those enjoying substantive autonomy, and between philanthropic private institutions organised through democratic control by academic staff (for example, Harvard, Oxford, Cambridge) and for-profit chains responding primarily to owners' decisions. There are advantages and disadvantages to each of these, with institutions enjoying state funding able to carve out more space for publicly beneficial activities, free from the demands of the market, but often less nimble in bringing change and with less structural flexibility. In marketised systems, and ones in which states do not provide all funding (the reality for most HEIs across the world), there are extremely difficult decisions to be taken in relation to the 'green premium', when 'paying more' for environmentally friendly options may involve spending less on educational quality or risking the financial viability of the institution. Furthermore, in the context of market competition for student recruitment, displaying green credentials may appear a win-win – bringing a positive environmental impact while at the same time enhancing the brand – but there is the constant danger of slipping into greenwashing.

In relation to those aspects directly under the purview of central management, leaders have control and influence over internal structures and external relationships (the latter the subject of the next chapter, Chapter 9). The task facing leaders is one of mitigation, adaptation and regeneration. Mitigation involves measures relating to Scope 1, 2 and 3 emissions, ensuring that they are measured as accurately as possible (acknowledging that there are challenges over Scope 3) and that steps are taken to reduce them – as outlined earlier in this chapter. As discussed in Alexander (2023), some adaptation measures for universities are essential, in protecting them from extreme weather, higher temperatures, flooding and scarcity (while at the same time not undermining the mitigation efforts). Finally, universities must actively regenerate, both within their own grounds and lands if they have them – through protection of biodiversity and forests, new tree planting, fostering of mutually supportive communities – and contributing to regeneration in adjoining communities.

However, as discussed through this chapter, most of the day-to-day activities of universities are relatively autonomous from central management, and need that level of distance and freedom for their proper functioning. The task for leadership therefore is to nurture climate action by creating the right kinds of conditions, and protecting those spaces from incursions and distortions. For example, student and staff time can be protected so as to enable engagement in climate-related activities; connections can be made between the curriculum and campus sustainability initiatives; incentives can be provided for staff through changes to recruitment and promotion policies, and to students through small grants and credits; and cross-disciplinary spaces can be created so students, academic and professional staff can debate and collaborate on climate-related interventions. In some cases, major institutional reforms can be important facilitators – for example the creation of interdisciplinary centres, departments and faculties, or the establishment of coordinated climate and sustainability hubs. Even mission statements and institutional pronouncements – while easy targets for cynics and susceptible to empty rhetoric – can play an important role in supporting and legitimising climate-related activity.

Importantly, the leadership and staff members of HEIs also need to be engaged in a process of learning themselves. As Sterling and Maxey (2013: 7) note, we need to think not only of 'designed learning' (the formal curricula provided for students), but also of the 'institutional learning' on which it is dependent:

> . . . the social and organisational learning that the policy-makers and providers may themselves undergo or experience: senior managers, academic staff, support staff, and policy-makers and stakeholders.

Finally, we need new leaders and innovators, outside of mainstream institutions, to create new kinds of institution. These new 'subversities' (Santos 2017) or 'ecoversities' (Mandel et al. 2022) need courageous vision and imagination as they may bear no resemblance to our contemporary institutions and will need to endure lack of recognition, incredulity and dismissal, in their task of refashioning higher education. Before exploring these new institutions in greater detail in the conclusion, the chapter that follows will assess the final dimension of university action, its relationship with external communities.

# 9
# Engaging with society

Dawn was breaking over the Indian Ocean, and the fishing boats of Somanga were being readied amidst the first shouts of friends and traders in the fresh morning air. A small group could just be seen wading through the warm salty water towards a paint-chipped wooden boat that would take them out to one of the coral reefs of the region – or what was left of it. They were going to what they affectionately called their 'farm', to plant, tend and monitor the reef, using an ingenious method of rooting fragments of coral in local bricks and allowing it to spread and regenerate. With few resources and only rudimentary equipment, community members have somehow managed to turn the tide of coral loss and regenerate the area.

The Tanzanian organisation promoting this work is not a government agency or international NGO. The rather unusual grassroots collective is made up of men and women from the local fishing community, concerned about the disappearance of the coral reefs they have known since their childhood, along with researchers from the University of Dar es Salaam, some way up the coast, who have come to learn from the community, systematise the knowledge generated and spread the ideas (Mazigo et al. 2023). This small partnership shows us what is possible in climate action, even in the most difficult circumstances, and faced with the day-to-day pressures of survival in an unequal world.

A few miles up the coast there is a rather different scene, in which the idyllic island of Songo Songo hosts the country's first gas field, including eight wells and a processing plant. Following the establishment of a Bachelor in Science in Petroleum Engineering at the University of Dar es Salaam, an increasing proportion of the professionals working here will come through the local flagship institution, eager to make something of their lives in the new economy. The oil and gas industry also relies on research and innovation and other partnerships established with universities, in an ecosystem of knowledge exchange for supposed economic prosperity.

How do we understand these apparently contradictory portraits? Is the university a force for positive change, galvanising social energy through its intellectual stimulus and collaborative creativity? Or is it bound up in hegemonic power, a handmaiden to the industrial model that has plunged us into climate crisis and keeps governments and peoples enthralled? These complex and multiple relations of universities with the society outside are thrown into stark relief in the context of the climate crisis.

This chapter will assess the implications of climate change for the relationships between universities and the external society. This area of work – commonly referred to as community engagement, but also going by a range of other terms – is a neglected one in relation to the 'core' activities of teaching and research, but is nevertheless gaining increasing prominence in the contemporary age and is particularly valued in relation to climate justice and adaptation to climatic shifts. It is also an area that presents particular difficulties for universities in practice, given the clashes of culture, practices and language between the academic community and other communities, the inappropriateness of conventional gauges of academic quality for assessing the effectiveness of this work, and the uncertainties over who should be funding these engagement activities in the context of financial constraints. From another perspective, these challenges are exactly what universities need in order to shake them out of their inert complacency, with community engagement presenting a precious opportunity for university transformation.

The pathways of climate impact framework put forward in Chapter 4 does not in fact use the term 'community engagement'. In part, this is because the scheme is categorising the activities that universities engage in – rather than those of other parties – and community engagement involves both. Another reason is the vagueness of the term. What goes by the name of community engagement can involve activities as diverse as providing free-of-charge Spanish classes for local residents, seconding academics to work in parliament, advising on policy, working with a local organisation to campaign against industrial pollution, and collecting data on changing weather patterns with nomadic pastoralists. These activities are united in involving joint work between those within and those outside universities, but constitute variously teaching, research, advocacy, provision of services and a range of other functions. Community engagement does not, therefore, provide an analytically tight category, but serves more as an aspirational call. This chapter will therefore use the term community engagement to indicate this broader aspirational area, while also assessing specific areas of activity defined as services and public debate.

Why is it necessary to consider external engagement in relation to universities? Would it not simply be in the normal run of things for an institution to engage with the society of which it is a part? Certainly, there is a form of engagement that any HEI undertakes while conducting its core activities: namely, the reception of students into the university to undergo a programme of study, and then their movement into the world once it is completed. There is also the gradual seeping out of academic writings, ideas and scientific discoveries into society, public opinion and technologies. However, these activities are not included in the common usage of the term community engagement; what is meant is those forms of engagement that are outside this core relationship. (The same, in fact, is true of many institutions – so a healthcare provider or law firm would not include treating patients and representing clients as part of their community engagement.) In some ways, then, what is included in the third pillar is by definition something that is out of the ordinary for universities, a distinct and deliberate way of engaging with the world outside.

The notion of community engagement raises questions about what is meant by community and what is meant by engagement – and the broadness of these terms allows for considerable contestation and a range of normative positions. 'Community' initially suggests the people living in the geographical vicinity of the university – sometimes referred to as 'local community' – but can also refer to other forms of community, such as associations, organisations and movements, and to geographical communities that may be distant, in some instances even companies and the government. The term community also assumes a unified body with a single voice – dynamics that rarely pertain in practice, and can serve to privilege certain voices and marginalise others. Engagement is equally elusive as a term. It can involve any form of contact, although requires something that is active, intentional and wholehearted.

Climate change places a spotlight on the community engagement role of universities. Universities (along with some non-university research institutes) are at the forefront of climate science but, as argued by Rapley et al. (2014), that science needs to filter through into the thinking, the practical work and the technologies of all parts of society. In part, this task stems from the social responsibility of universities in supporting local communities that do not share in the privileges of the university community and may be particularly vulnerable to climate impacts. But further to these disparities, climate action requires various forms of collaboration across boundaries. It needs multisectoral working across different sectors, combining diverse areas of professional work; it needs multi-, inter- and transdisciplinary working across different academic areas; it needs a dialogue of knowledges bringing together different worldviews and epistemologies; and it needs co-production of research, involving different kinds of actors. The sections that follow will explore further how these forms of co-production may take place. We have some effective methodologies available and some inspiring cases, although it remains a highly challenging form of work.

This chapter then will explore the possibilities and challenges for HEIs of contributing to the important task of knowledge production, exchange and distribution together with external communities, in light of the specificities and urgency of the climate crisis. It starts with a brief exploration of the evolving role of external engagement in higher education historically, before moving on to consideration of the service provision and public debate roles specifically. The chapter then puts forward an analytical framework for understanding the various dynamics involved, along with some normative considerations for universities.

## Evolution of the third pillar

It is well known that universities have struggled historically in developing close and collaborative relationships with the communities that surround them. The various images associated with the university – the 'ivory tower', the 'gown' separating it from the 'town' – point to a self-consciously and preciously separate existence unwilling to sully itself in the world outside. The links between universities and monasteries – in some cases, they were one and the same – have encouraged this perception. In practice, there has never been a historical period in which universities were entirely isolated from the rest of society: they have always been subject to the worldly aims of those funding them (whether patron, institution or student), and connected to the world of work through their preparation of lawyers, clergy, doctors and, over the centuries, an increasing range of other professionals.

However, in the nineteenth and twentieth centuries a much more conscious and transformative opening up to society occurred. The economic role of higher education became increasingly recognised, and with it the need to align its programmes more closely with the changing nature of the labour market, leading to the inclusion of engineering, agriculture, accountancy, nursing and teacher education, among many other areas. Simultaneously, there emerged a social justice mission, which sought to use universities for alleviating poverty, fostering democratisation and promoting equality. These dual tendencies manifested themselves in different ways around the world. In the USA, the Morrill Act of 1862 led to the creation of so-called 'land-grant'[1] universities, with an explicit commitment to engaging with local agriculture and industry (McDowell 2003). These institutions exist to this day and have not only expanded access to higher education but also developed extensive programmes of outreach – for example, in the area of agricultural extension – and have been highly influential in national development. In Latin America, the 1918 reforms in the National University of Córdoba, Argentina, led to a region-wide transformation of public universities and the emergence of a strong social justice commitment, again with expansion of access and strengthening of community links, but in this case also introducing democratic processes within university governance. In Africa, the model of the 'developmental university' emerged in the post-independence period in the 1960s and 1970s, with national universities taking on an explicit orientation towards promoting the interests of the poor and the marginalised, and rejecting the elitist Eurocentric models that had been inherited.

These and other movements changed both the self-conception and the practices of universities in adding a third pillar alongside teaching and research – admittedly with much less importance than the other two, but at least with a place at the table. However, this third pillar is highly diverse, and contains within it a range of orientations and types of activity. In addition to the social justice-oriented activities associated with the developmental university, this third pillar also includes entrepreneurial ventures, relating to what is sometimes called 'third stream' activities (as in a stream of income). In this case, there is engagement with external communities not in order to contribute to them but to extract revenue to support university finances. In some cases, these activities may involve the setting up of spin-off companies drawing on technologies and basic research developed within the university, thereby generating revenue for the university. There may also be knock-on positive impacts for the economy as a whole. This vision is supported by the triple helix model, in which government, university and industry operate in a mutually supporting relationship (Etzkowitz and Leydesdorff 1995). In marketised systems of higher education, such as the USA and UK, it is this form of third pillar that predominates.

It is not impossible that these kinds of activities will play a role in addressing the climate crisis. Bill Gates' book *How to Avoid a Climate Disaster* (2021) sees this model of innovative, science-based start-ups as the answer to the challenge of moving the world to a carbon-free existence, in developing energy sources and products that will be effective and cheap enough to wean people off fossil fuels. Certainly, technological innovations will be very helpful in addressing climate change if we can make them generally available to the global population. Nevertheless, as argued in the first part of this book, it will never be the entirety of the solution, as the problems of unsustainability are rooted much more deeply in our economy, politics and our very selves. Furthermore, an income-generating approach taken by universities will never achieve the kind of broad engagement of different stakeholders needed to coproduce knowledge. Universities, therefore, need to engage with external communities in ways that see them not as consumers of their products, but as partners in addressing the climate crisis, in a reciprocal and mutually beneficial relationship.

In fact, theorists of the triple helix have now added two further helices. The quadruple helix moves from knowledge economy to knowledge society, adding a fourth element of 'media-based and culture-based public' or 'civil society' (Carayannis and Campbell 2009). The quintuple helix brings in the natural environment, with sustainable development brought about through the circulation of knowledge, know-how and innovation between five helices (educational, economic, environmental, public and political) (Carayannis et al. 2012). Engagement with the ecological is certainly very welcome in this new framework and puts a spotlight on the absence of the natural environment in the dominant triple helix. However, all these models are highly economistic, based on the human capital model of education, and assuming that economic growth in a capitalist system is both desirable and inevitable.

The array of different terms used in relation to university engagement with society have quite different connotations, although not always aligning with the actual practices. One of the earliest terms is 'extension', used commonly in relation to agriculture, and often in Latin America (in its Portuguese [*extensão*] and Spanish [*extensión*] versions). This term suggests the movement of knowledge from university to the outside, in a mono-directional relationship that brings benefit to the receiver. So, the agricultural research produced in the university student is transmitted via an extension agent to a farming community, which then increases its yields. 'Outreach' also has these kinds of connotations, showing the privileged centre (university) extending its largesse to the less fortunate periphery. However, it is important to acknowledge that work carried out under these labels does not always correspond to these connotations: for example, extension is used with a collaborative perspective in Brugger and Crimmins (2015).

The term 'community engagement', on the other hand, opens the door to a two-way flow, and to co-production as well as knowledge and tech transfer. Another term used is 'public engagement', which can be equivalent to community engagement but will often focus more on the public debate and awareness-raising aspects. In North America, the terms 'service' and 'public service' are common, indicating a positive contribution to society but one based on the perceived superiority of the university and a one-way flow of influence. In Europe, the catch-all term 'third mission' is common. As mentioned above, 'third stream' activities are often those of a commercial nature and in some universities they are simply called 'enterprise': the latter term, while potentially including social enterprise, is normally reserved for activities of financial benefit to the institution. There are also a range of terms that focus more specifically on movement of knowledge, with 'tech transfer' and 'knowledge transfer' indicating the more traditional one-way flow, while the more recent 'knowledge exchange', as used in the UK's Knowledge Exchange Framework, is more collaborative.

The second half of the twentieth century saw various movements towards substantive democracy (rather than mere formal entitlements), which manifested in the general political sphere through participatory democracy and direct democracy, in the workplace through cooperatives and worker control, and in international development through participatory rural appraisal and participatory action learning, among other approaches (Chambers 1994; Pateman 1970). These ideas have also influenced community engagement, with the rejection of approaches that treat local communities as objects to be acted on, but instead involve community members as active participants in the process. Taken further, this trajectory has allowed the emergence of co-production, of community members being agents of research rather than just beneficiaries, and the use of participatory action research, through which they work together with university-based researchers to transform their communities. These approaches will be discussed in greater detail towards the end of this chapter.

We can observe, therefore, a general trajectory over the centuries towards stronger engagement between universities and external communities, and a greater range of forms of engagement, although also a diversification of intentions, variously approaching engagement as a form of charity to less advantaged communities, or an act of responsibility, social justice or self-interested financial gain. The sections that follow will provide examples of the two categories of engagement included in the framework presented in Chapter 4 (services and public debate), followed by the presentation of a framework that maps this diversity of form and intention of engagement.

## Services

If universities' engagement with communities were reduced purely to services, it would certainly be inadequate. Services indicate a one-way flow, without a strong collaboration with external partners, and assuming that the knowledge of value is held within the university. Nevertheless, services can be vital when there are particular needs within community or society that universities can fulfil, and as a first step towards more collaborative partnership.

What is it in reality that universities can offer by way of services? Universities' core strength is knowledge, both embodied and disembodied. Its staff (and often students) are specialists in research and analysis across the full range of areas of knowledge, and can apply that expertise to the specific problems facing external organisations. Consultancy, technical advice and secondments of staff are therefore important ways in which the university can contribute to climate action. Naturally, there can be a considerable financial cost to diverting staff from their normal functions, meaning that very often these formal arrangements are made with well-funded organisations, in particular public authorities and businesses. In cases in which university staff have a strong personal commitment, they will often lend their services at low or no cost.

Yet there is also disembodied knowledge, in the form of research findings, theories and innovations that are developed within universities and can be applied outside. Most often this knowledge is codified and distributed in the form of journal articles and other academic publications, which are (journal paywalls notwithstanding) generally available as a public good. In some cases, there are intellectual property restrictions, and institutions will make the knowledge available on a commercial basis.

Beyond these two core elements, universities also have organisational capacity along with physical and virtual infrastructure and estates which can be deployed to provide services. So, for example, rooms and lecture halls can be provided for meetings and events, equipment can be provided for environmental monitoring purposes, and the administrative capacity of the university can be utilised for organisational purposes. Very often all of these elements are used together. So, for example, one of the most common and impactful services provided by universities are health clinics and hospitals. In these, university estates and buildings host the service, but embodied knowledge is also deployed in the form of health professionals, as well as disembodied knowledge in the form of medical research.

There are many examples of services of this type provided specifically in relation to climate. Hillmer-Pegram et al. (2012), for example, analyse an innovative greenhouse gas emission inventory carried out by researchers from Penn State University for the local government of Pennsylvania Central region, leading to a collaborative climate change mitigation plan being developed. Wesselink and Gouldson (2014) present the case of the 'mini-Sterne review' and collaborations between academics and Leeds City Council to analyse the economic implications of transition to a low-carbon economy at the local level. Booth et al. (2020) report on a course on carbon and energy management for local businesses offered by the University of Northern British Columbia, an initiative that also involves the university students providing advice for the small and medium-sized companies on how to reduce fossil fuel usage. The extent and nature of this service provision differs markedly by context. In some contexts (particularly in cases in which there is lower coverage of higher education across the population), there are close relationships between universities and different levels of government, with university professors regularly moving into secondments and permanent government roles. In others, relationships need to be more consciously constructed through policy and funding levers. Collaboration with the private sector also varies greatly, depending on the financial incentives for the university (strong in marketised systems in which government funding has receded), and also on attitudes towards business on the part of university researchers, and the extent to which critique and opposition to corporate misdemeanours, or alternatively constructive engagement, takes the upper hand.

While it is usually assumed that staff members are involved in these forms of community engagement, it can as easily be students. In fact, participation of students is an important form of transmodality, with opportunities for both positive impact on communities and student learning (Mtawa 2019). The University for Development Studies in northern Ghana, for example, has a 'third trimester programme' through which all undergraduate students spend a period in a rural community each year, carrying out a collaborative development project and learning from community members. In Brazil, this kind of experience has been formalised through the 'curricularisation of extension' policy, through which all undergraduate students now need to have 10 per cent of their curricular time involved in community engagement work.

Another specific area here is in relation to schools. Universities have a particularly strong contribution to make in relation to the lower levels of the education system. They do this primarily through teacher education, being the main provider of pre-service and in-service training, but also through transposition and translation of bodies of academic knowledge into the school curriculum, and other direct contributions to enhancing quality. Specifically in relation to climate change, universities have an important role to play in supporting schoolteachers with the technical knowledge needed to raise awareness of climate among students, and in providing spaces within the university for teachers to develop their practice. In most contexts, coverage of climate change in teacher education needs to be significantly improved. A survey carried out with schoolteachers in the UK by UCL's Centre for Climate Change and Sustainability Education (Greer et al. 2023) showed that most relied on their own initiative and information-gathering, with only 13 per cent having engaged with issues of sustainability and climate in their initial teacher education, and 45 per cent in formal professional development.

There are also direct connections between universities and schools. Reimers (2021) argues that there is a win-win in universities engaging with schools as it provides university expertise to support curriculum development but also a crucial learning experience for the university students. Asherman et al. (2016) describe an initiative of this type with MA students presenting greenhouse gas demonstrations to primary and high school students in France, while Monroe and Oxarart (2019) look at contributions to high school biology in southeast USA, involving collaboration between researchers, students and school teachers to develop high-quality materials.

## Public debate

The dimension of public debate has, generally speaking, had the least attention of all the functions of the university. In part, the reason is that it is often undertaken by staff in an individual capacity and in spontaneous ways, and as such is outside the planned official functions of the institution. Furthermore, there has also been some scepticism as to whether it really is within the purview of universities to be influencing public opinion – that is, beyond conventional dissemination of impartial research findings.

Nevertheless, in relation to climate change this dimension has been highly visible. University academics have been prominent voices in alerting the public to the reality and catastrophic risks of climate change, advocating for immediate and far-reaching action and battling with sceptics and deniers. Students have spearheaded campaigns and protests both within and outside their institutions. As discussed in Chapter 8, there is a difficult question as to whether and in what circumstances universities should take a substantive position on questions of value. Yet in fact most of the public engagement in relation to climate change is taking place independently of unified institutional positions. Academic freedom and the unusual 'loose coupling' (Clark 1983) of the university as an institution has meant that employees are allowed to express their personal views, and do not need to speak on behalf of their institutions, even in public events. (There are some exceptions here – for example, those in positions of senior leadership.)

Contribution to public debate can take a number of different forms. It can stem from a 'leaking' of outputs primarily intended for the academic community (journal articles, academic books, conference presentations) to the general public. In order to maximise this possibility – and also to ensure a more level playing field across academic communities in different parts of the world – there has been considerable attention to open access to academic publications in recent years (Fyfe et al. 2017). Yet engagement very often takes place in more active formats through participation in traditional media (television, radio, newspapers) as invited 'experts' or in a more journalistic role, and in new social and digital media (blogs, websites, Twitter/X). Kukkonen et al.'s (2020) study of contributions to newspaper media debates on climate change in the Arctic in Finland and Canada shows that universities are among the most prominent voices, along with governments (more present even than environmental NGOs), although, tellingly, more internally pluralist than the other organisations in terms of messages.

Students are also present as voices on both traditional and social media. Yet their influence in relation to climate change has been most prominent in mobilisation, campaigning and protest. While secondary school students have led the best-known campaign in this regard (Greta Thunberg's Fridays for Future), university students have also been highly active. One prominent area of protest has been fossil fuel divestment, through which universities remove from their investment portfolios those stocks most directly contributing to climate change. According to Grady-Benson and Sarathy (2016), at that time there were 400 divestment campaigns in the USA and 560 internationally. The authors discuss the transition in these movements from individualised efforts to building a collective movement. Many institutions have rejected divestment on the basis of various arguments such as their fiduciary responsibility for investments, the limited impact of divesting from fossil fuel companies, and the greater influence exerted on companies by remaining a shareholder. Nevertheless, there have been a number of notable successes (for example, Harvard University), and while others have rejected divestment, they have nevertheless (and perhaps as a response) implemented other carbon reduction measures. Interestingly, students in the study affirmed that the relevance of their actions was not so much about the direct impact of reducing investment in stock as stigmatising the companies in question – making it more of a public relations exercise than a financial tool.

Students also participate in protests, civil disobedience and public disruption. In 2023, climate activists occupied universities in Belgium, Germany, Portugal, Spain and the UK (Gayle 2023). A student protest the previous year at the University of Barcelona convinced the university authorities to introduce a mandatory climate crisis module for all undergraduate and postgraduate students, as well as staff training (Burgen 2022). In November 2023, Just Stop Oil activists sprayed Oxford University's Radcliffe Camera with orange paint (Haskins and Prideaux 2023). As with all protests of this type, the activists gamble with the dual impacts of obtaining widespread media coverage, while at the same time producing negative reactions from certain segments of the public (Mann 2022).

The above examples have covered the various ways in which university staff and students engage in debates in the wider society. However, there is a second role that universities play, that of providing an arena for public debate. This role acknowledges the fact that – in spite of the incontrovertible reality of anthropogenic climate change – there are still a number of unresolved questions, particularly relating to how to bring about the necessary societal transformation so as to live within the limits of the ecosystem. Universities play an important role by providing a space for vigorous and well-informed debate, either a literal space on their campuses through events, workshops and debates, or a symbolic one through their promotion of dialogue in other societal spheres. Inevitably, this role raises questions of what might be considered acceptable in terms of free speech, with universities in recent years a battleground between those defending absolute academic freedom and those mobilising around 'no-platforming', preventing expression of certain views (for example, racist, fascist).[2]

There have been some criticisms of how universities have engaged in public debate in the broader society. George Marshall (2014) highlights academics' inexperience and naïveté in playing the media game, and Rapley et al. (2014: 8) concluded that:

> Climate scientists are finding themselves ill-prepared to engage with the often emotionally, politically and ideologically charged public discourse on the evaluation and use of their science.

At the same time, climate scientists have showed courage and resolve in the face of threats and intimidation (Ley 2018; Polsky 2019): one of the messages received by Michael Mann, for example, requested that he be 'shot, quartered and fed to the pigs, along with your family' (Marshall 2014).

The public debate role is one of the most complex, as naturally those working and studying in universities hold widely diverging opinions on the matter, and it would be both impossible and inappropriate to ensure consensus of messaging. So, is there any kind of absolute good here in relation to public debate? It can be argued that simply raising awareness of climate change and providing an arena for debate are important, even if a variety of different positions on it are promoted. Inevitably, the practical expressions of this principle vary markedly between countries – and may be highly restricted in authoritarian states – and between higher education systems with their varying regulatory frameworks and cultures.

## A framework of university engagement with society

As stated above, the triple helix model (and even its quadruple and quintuple reincarnations) are characterised by an economistic conception of the university and indeed of the goals of society. Yet there is another limitation in that the schemes assert the existence of interaction between the different parties but, beyond the circulation of knowledge, say little about *how* these things happen. To address this gap, this section puts forward a framework for mapping the various dynamics and components involved in those interactions.

There are six main dynamics when it comes to universities' engagement with external communities. The first is *porosity*, the extent to which there is an open border between universities and society, allowing for movement in and out. This idea was explored in a previous volume (McCowan 2019) in relation to the 'interaction' dimension of the university, with the potential for movement of both ideas (through publications, media, events) and actors (through secondments, projects). Universities have a semi-permeable membrane with considerable variation between institutions, and levels of porosity may differ depending on whether the movement is *inbound* or *outbound*. While generally speaking porosity is considered a positive quality of HEIs, there are dangers of excessive openness ('hyperporosity') in which the institution ceases to have a distinctive space and its own identity and protagonism.

Second is *flow*. The movement in question can either be inwards or outwards, with ideas, practices and actors moving from the university out into communities and society, or alternatively those influences coming from the outside in. As discussed above, traditionally universities have seen themselves in a superior position, holding the truths and techniques of value, and sharing them when time permitted with the less privileged groups outside. However, this attitude of largesse has been challenged in recent years, with greater acknowledgement of the knowledge production capacity of non-university communities, the need for their collaboration in the application of knowledge, and of the benefits to the university of absorbing ideas from outside. These dynamics are particularly relevant if we value epistemic pluralism and the creation of an ecology of knowledges. And they are also particularly relevant in the context of climate change, which requires multisectoral working involving different forms of actor.

The third element is precisely this question of the various forms of community to be engaged with – the *actors*. There is huge diversity possible in this respect. Universities can engage openly with the whole of society – for example, through the dissemination of ideas in social media, mainstream media, books and other outlets, or the creation of technological innovations that are made generally available on the market. Alternatively, engagement can be targeted towards specific organisations, whether the government and public authorities, businesses, NGOs or community organisations. Finally, there are possibilities of close engagement with individuals. As discussed in Chapters 4 and 5, there are questions both of scale and intensity of the engagement here: sometimes engagement works across a broad range of actors with a low intensity of impact (for example, in the case of a tweet raising awareness of climate dangers) or alternatively with a small number of actors with high intensity (for example, in a three-year participatory action research project with a small group of community representatives).

In this dimension we can also consider who the actors are within the university: whether it is academic staff, professional staff, students or some combination. The different configurations of internal and external actors will bring different benefits and outcomes on both sides. One of the dynamics of interest here on the part of the university is the level of isolation/integration with other activities (teaching, research and campus operations). If they are highly integrated – displaying 'transmodality', or being 'embedded' in the terms of Farnell et al. (2020) – then there is greater potential for transformation more broadly.

Fourth, there is variation in relation to the *actions* in question. According to the pathways to climate impact framework presented in Chapter 4, these broadly take the form of services and public debate. Within these categories there is a wide variety of different kinds of activity, with services consisting of, for example, running health and legal clinics for disadvantaged populations, conducting consultations on local environmental issues, creating spin-off firms, partnering with businesses, sending technical experts to work in government departments and so forth. Actions will normally be located outside of the space of the university, but can also occur inside (for example in the case of a public event or workshop).

Next there is the *orientation* of the action – that is, its purpose, motivation or intention. As discussed above, these orientations can vary widely between benefiting society or the institution, or both, and are built on a wide range of different value bases. They do not necessarily determine the outcomes of the work but will naturally influence the way in which it is conducted and the likely outcomes. Orientations can include financial gain (creation of a spin-off company), intercultural understanding (student sojourns in a rural village) or poverty reduction (agricultural extension). Motivation may differ between the university and community sides, and in some cases (such as seeping flows of ideas) there may be little intention at all. Connected with this point is the attitudes to the other on each side of the university/community boundary (for example, trust, respect), which can influence and indeed determine the effectiveness of collaborative work.

Finally, there is the *level of collaboration* between university and external community. 'Third pillar' activities necessarily involve some kind of contact between university and community. Yet they can take place in ways in which each party acts in relative isolation from the other, or without sharing decision-making. So, for example, a workshop on monitoring pollution levels in a local river may be provided for community members within the university but designed entirely by university staff without any consultation. Or, alternatively, the university may fund the construction of a greenhouse in a local village to act as a seed nursery for reforestation, but the management of the nursery may be carried out solely by the community. Co-production, naturally, involves a different kind of interaction, one in which university and community act in concert, sharing decision-making, action and analysis. These six dynamics are represented in Figure 9.1.

The diagram below shows both the 'what' and the 'how' of universities' engagement with society. In the first case, it shows the range of substantive foci (for example, environmental monitoring or radio debate), the motivations or orienting principles, and the kinds of actor involved (students, local farmers and so forth). Only a suggested list for each category is provided – there are a huge range of different possible actions. It also shows the dynamics of how that engagement takes place: the direction of flow of influence between university and community, the level of porosity (or the 'ivory-tower-ness' of the university) and the extent

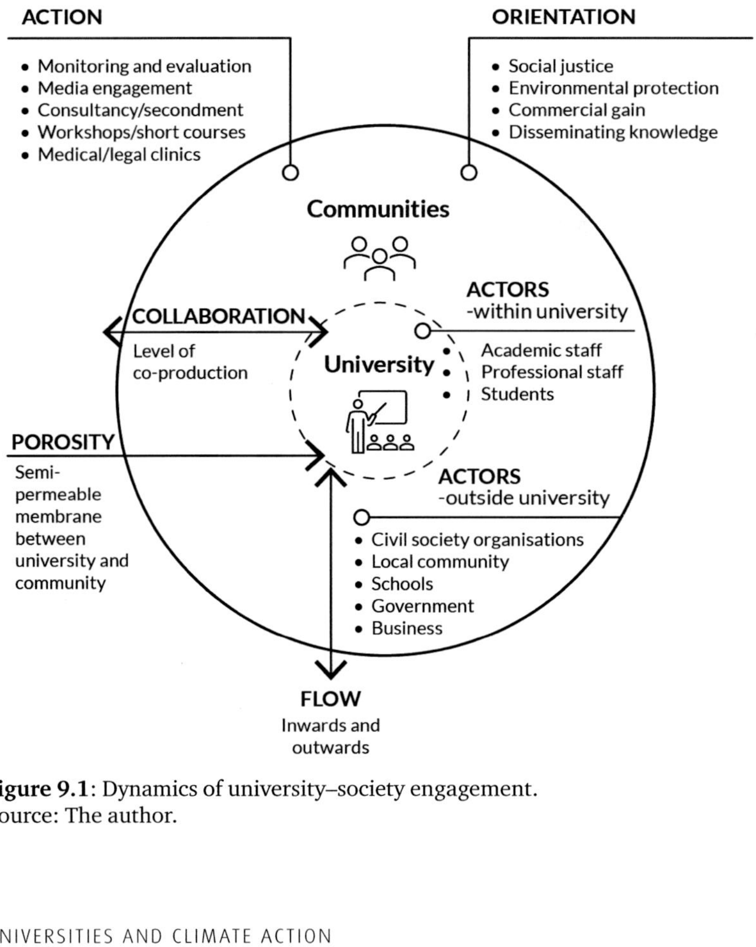


**Figure 9.1**: Dynamics of university–society engagement.
Source: The author.

of co-production (to be explored further in the section follows). These dynamics will influence the nature of the engagement work, but also the extent that each gets changed by the other and benefits the other. While universities have traditionally thought about their positive influences on society, it is important to acknowledge both the possibility and indeed the necessity of external communities transforming universities.

Any visual representation of the relationship between universities and society will necessarily be reductive. Some universities have a single physical campus with a boundary clearly demarcating what is inside and what is outside the institution. Yet even in these cases, the representation cannot capture the fact that most of those working within that boundary live outside. So university staff and students are very often also residents in a local community, and as residents may have interests that are different from those of the university. Many universities operate across multiple campuses, including branch campuses in different countries. Furthermore, the institution of university is not only its physical location, with the activities of its staff and engagement with its ideas and technologies occurring in various places – in some cases globally. Developments in information and communications technology and the rise of online learning have, of course, intensified this dynamic, with physical campuses having an increasingly fragile claim to identification with the institution.[3]

Nevertheless, visual representation can still be useful, as long as we bear these caveats in mind. The boundaries and locations of the different actors are intended to provide a visual support for conceptualising the relationships, rather than representing the literal location of each. One other element that is important to bear in mind is that the boundary between university and community is traversed by individuals and organisations that serve as bridges – sometimes explicitly for this end. Brugger and Crimmins (2015: 23) discuss boundary organisations, which are 'situated at the boundary between science and society with accountability to both' and:

> . . . engage in the coproduction of usable local-level adaptation strategies through the collaborative participation of actors from both sides of the boundary and professionals from the organization who serve a mediating role. (Brugger and Crimmins 2015: 23)

They present a case study of one boundary organisation, the US Cooperative Extension System, 'with the aim of bringing research and educational programs developed at universities and agricultural research stations to the broader public' (Brugger and Crimmins 2015: 25). In addition to boundary organisations, there are also:

> . . . information brokers – intermediaries between knowledge producers and users who are fluent in both worlds; and knowledge networks – informal networks of knowledge producers and users who share information across areas of practice. (Brugger and Crimmins 2015: 22)

There is, therefore, not a simple membrane between university and community, but a border area that is mediated by various organisations and individuals. Furthermore, and most extensively, there are the graduates of university who in their later lives maintain connections to their alma mater. (The role of bridging actor is represented in the pathways to climate impact framework presented in Chapter 4.)

What are the normative implications of this framework? If we plot activities of the university on this graph, would we want them to be clustered in particular areas? Some of the questions would certainly seem to have clear normative implications. In most instances, a high level of collaboration between partners would be desirable, rather than working in isolation, in order to reap the rewards of diverse perspectives and practices, and also to ensure the needed sense of ownership on all sides. Greater porosity would generally be seen to be a good thing, although with the caveat mentioned above that some distinctive space for the university is important. In terms of orientation, there might be a range of legitimate motivations, while preferably avoiding more self-interested ones (for example, purely financial interest, or exploitative or extractive intentions) or ones actively contrary to climate action. In most cases, however, the primary value is diversity – to ensure the greatest possible range of different actors, forms of engagement, and inward and outward flow. One area, however, in which there are clear normative implications is in the level of collaboration, and the possibilities of co-production.

## Towards co-production

As indicated on Figure 9.1, it is possible for universities to engage with external communities in relatively arm's length ways. Universities can provide a service for the general public and community members may attend a course, but without any substantive working together in the creation of the ideas and realisation of the practices. Attention, therefore, is needed to the level and nature of collaborative working. Central to this debate in recent years has been the notion of 'co-production', the idea that research, and the development and application of knowledge, can be carried out jointly between universities and communities, drawing on the distinctive strengths of each.

The notion of co-production draws on a long tradition of efforts to bring research and scholarship out of the ivory tower and in contact with society outside, as discussed in relation to the developmental university model. Brugger and Crimmins (2015: 21) discuss the notion of 'usable science', which:

> . . . responds to the recognition that the traditional 'mode 1,' 'linear,' or 'loading dock' model of science, based on a strict separation of science and society and a one-way flow of knowledge between them often, results in information that is not useful to decision makers.

Furthermore:

> It is informed by the understanding that the boundary between science and society is mutually constructed and proposes that a two-way flow of information between knowledge producers and users will result in the coproduction of knowledge that is perceived as salient, credible, and legitimate by relevant stakeholders or decision makers and will therefore be more effective in influencing social responses. (Brugger and Crimmins 2015: 21)

The bringing together of different actors inevitably encourages interdisciplinary working, as well as an ecology of knowledges. As emphasised by Binagwaho et al. (2022), the task at hand is partly to recognise different ways of knowing, but also knowing through difference:

> When the circumscribed 'mainstream' noun-like categorical model of knowledge is expanded into this more processual and relational (verb-like) understanding of knowing. (Binagwaho et al. 2022: 46)

Trencher et al. (2014) apply these ideas to environmental regeneration through the notion of 'co-creation for sustainability', a new function of universities to take the place of the reductive economic third mission, and its focus on one-way technology transfer. The case of Oberlin is presented as a notable example of regeneration of the city and local area through partnership between university and community.

Binagwaho et al. (2022) provide a number of other examples of successful co-production, including in the development of the Danish wind turbine industry through the Folkehøjskoler (folk high schools), as well as work between universities and Indigenous peoples in the North American Arctic.

These ideas link strongly with the discussions of Indigenous knowledge and epistemic pluralism in Chapter 3, and the need to incorporate diverse knowledge traditions in research agendas in Chapter 8. Indigenous institutions have a particularly important bridging role in this regard, as seen in the intercultural universities in Mexico (Perales and McCowan 2021), and in the 'tribal colleges and universities' on reservations in the USA (Fillmore et al. 2018). Nevertheless, there is still a significant lack of incorporation of Indigenous perspectives in mainstream research, as in the case of the IPCC reports (Ford et al. 2016). When their perspectives are included, it is often in partial and largely instrumentalised ways (García-del-Amo et al. 2020). The term 'Indigenous' is mainly applicable in contexts in which there is a distinct settler community (for example, the Americas, Australia). Yet in many other contexts there are traditional rural populations that may retain forms of knowledge distinct from the mainstream, in equivalent ways, so it is important to acknowledge and include these other forms of local and traditional knowledge – as is the case with the non-Indigenous communities in the Amazon rainforest (Hage et al. 2023).

Participatory action research is one of the primary methodologies for co-production. Originating in Latin America through the work of Orlando Fals Borda, drawing on the earlier ideas of Paulo Freire (1970) on radical shifts in the teacher–student relationship towards horizontal dialogue, it advocates for non-extractive research relationships in which community members go from being objects or beneficiaries to active participants. Through cycles of reflection and action, multi-stakeholder groups collaboratively design and implement interventions, transforming themselves as well as bringing concrete changes in their environments. The Climate-U project is an example of participatory action research in relation to climate change, in which 15 institutions in six countries carried out diverse interventions engaging with marginalised actors in the research process. The synthesis of learning across these diverse experiences generated the following key ingredients for effective practice – equitable partnerships, co-production, immersion, agency and transformative institutions – in a cyclical and mutually reinforcing relationship (Climate-U 2023).

Co-production involves not only collaboration between groups of people and inclusion of marginalised actors but also the meeting of different forms of knowledge. The idea of ecology of knowledges and epistemic pluralism discussed in previous chapters is relevant, therefore, not only to the internal workings of the university (the curriculum and research processes) but also to its interaction with external communities. Community engagement represents an ideal opportunity for dialogue between mainstream academic knowledge and local or Indigenous knowledge, and in some circumstances creation of new forms from their encounter.

## The limits of community engagement

Most conceptions of the university now give recognition to a third pillar alongside teaching and research – variously known as public service, community engagement or extension. Yet what goes by the name of engagement with external communities varies dramatically, both in its extent and nature. Addressing climate change involves in the first place expanding the scope and increasing the visibility of this engagement, in light of competing objectives for the institution. Furthermore, a transformation is needed in the ways in which universities and communities interact. In relation to Figure 9.1, this can be understood as a greater porosity between university and community, not only in terms of the traditional task of 'disseminating' knowledge but also as a way of opening itself up to the ideas and priorities of those outside. Beyond the question of flow, there also needs to be a transformation in the space of collaboration, in ensuring that universities and communities work closely together, not only to apply knowledge in diverse contexts but also to coproduce it.

All aspects of the transformation needed in universities present their challenges, but community engagement is perhaps the most difficult to achieve. It is important to recognise that co-production is an ideal, but is rarely attained fully in practice, even when there is sincere motivation to work collaboratively. Barriers include hierarchies of power, lack of confidence and lack of time on the part of community members, and lack of cultural sensitivity and practical knowledge on the part of university staff.

Despite the lip service paid to public engagement, and economic rationales such as the triple helix, global higher education policy and norms do little to encourage it in practice. The idea of the 'world-class university' that is the aspiration for most research-intensive institutions has little space for community engagement, and the international rankings rarely include any indicator relating to it. In part, this absence reflects the low value placed on the activity, but there are also barriers presented by the difficulties of measurement, with gauges often as paltry as the 'number' of community engagement projects.[4] In the context of marketisation, there are also significant financial barriers: while third stream activities have an important place in the concept of the entrepreneurial university, they are restricted to the creation of products with profit potential in the market, leading to neglect of public goods and of engagement with lower-income communities that do not have sufficient wealth to purchase the university's services.

Furthermore, even when universities can carve out space for this work, there are barriers to the kind of equitable, coproductive engagement that climate change requires. There are disparities of power – in favour of universities in the case of local communities, but not in all cases (for example, government) – that hinder collaborative working. There are epistemic clashes, different ways of knowing, acquiring and communicating knowledge, different terminology and language used, as well as the kinds of contestation seen also in classrooms. There are also questions of durability and continuity of the work, with a predominance of short-term and fragmented projects. While this chapter has viewed the process primarily from the university perspective, there are also issues within communities, grappling like all human societies with preconceptions, hierarchies and discriminatory practice.

Finally, there are some limits on how much opening to society is actually desirable. As argued in a previous volume (McCowan 2019), some 'insulation' is important for universities in maintaining a distinctive character and a space for deep and sustained reflection and scholarship. In most cases, universities need to open themselves more to society. But in cases of 'hyperporosity' in which the university becomes indistinguishable from what surrounds it, it becomes simply another agency, NGO or company, able only to implement, but not to think, design, create and imagine. There is a temptation to assert that university should be doing everything and anything; it should reach beyond its boundaries but it cannot escape its fundamental nature, and community engagement has to be linked to the processes of open-ended enquiry.

## Notes

1 A problematic term from the perspective of displaced Indigenous peoples.
2 In response, the UK government passed the Higher Education (Freedom of Speech) Act in 2023.
3 Bacevic (2019) goes further to describe universities as *assemblages*, bundles of elements changing over time and without a clear boundary.
4 The Towards a European Framework for Community Engagement in Higher Education (TEFCE) Toolbox (Farnell et al. 2020) represents a significant step forward in mapping and researching universities' engagement with society.

# 10
# The road ahead

The character Crake in Margaret Atwood's novel *Oryx and Crake* (2003) lets loose a devastating virus that wipes out mankind, leaving just a small band of new humans, genetically adapted to the new climate reality. In the Marvel film *Avengers: Infinity War*, Thanos snaps his fingers and instantaneously wipes out half of the universe's population, easing the pressures on the ecosystem and allowing for the survival of those remaining. Both events echo the Old Testament flood story in which – albeit for moral rather than climatic reasons – the people of the Earth are drowned so that Noah and his offspring can begin afresh.

Although perhaps less colourful and dramatic, the responses to climate change that will be needed if we do not take action will ultimately be as devastating as these narratives. Major resource depletion, catastrophic weather events and resulting conflict will open the door to authoritarian governments that will put in place whatever measures are needed to protect at least a portion of humanity from annihilation. However inconvenient the shifts required to transition to a low-carbon society may be, they hardly compare with what awaits us if we do nothing.

The alternative is transformation from within. Individual and collective change is needed so that people can understand the roots of the crisis, identify the required action, adapt their behaviour and transform the structures of society. Ultimately, this kind of change is more effective than any authoritarian measure, as it puts in place the conditions for a lasting alternative. Forced measures of control and deprivation may bring temporary impact but will revert back to unsustainable forms of living if people do not themselves endorse the new way of life (parallels here can be seen with the flimsiness of political revolutions brought about through military victory led by a small avant-garde, without a deeper societal transformation).

Education is an essential part of this transformation from within. It underpins both the knowledge and understanding necessary to engage with this complex crisis, the communicational ability to deliberate and find collective solutions, and the emotional awareness and shifts in values to move from knowledge to action. Clearly, not any education will fulfil this task. The forms of formal education that have dominated around the world have been successful in generating literacy and numeracy on a mass scale, along with the sophisticated scientific and professional preparation of an elite, allowing for the rapid development of industrial society, mass consumption and exploitation of natural resources. Education historically has been, at the very least, complicit with the climate crisis, if not solely responsible for it. But the global experience of schooling has not been all bad. Simultaneously, it has also opened the door to critique and questioning, creativity and innovation, all of which will be essential if we are to revert the tide. Rather than abandoning altogether the formal structures of education that we have developed, we need, then, to find within them the seeds of a regenerative system of learning for sustainable futures.

This book has put forward two principal arguments: first, that higher education is essential to effective climate action at the global level, and second, that in order to fulfil that role, transformation in higher education is needed. Climate change as a crisis for humanity has certain key characteristics: it is anthropogenic, being rooted in human societies, politics, economics and culture, as well as psychological make-up; it is complex, in resisting linear, predictable solutions; it is contested, giving rise not only to denial and resistance but also legitimate differences of perspective on responses; it is time-bound, with delays in action progressively increasing the difficulty of responding; and finally, it is global, affecting all and therefore requiring collective solutions but disproportionately distributed in terms of responsibility and impacts. These characteristics make education as a whole essential, but specifically the university, with its bringing together not only of the educational intervention, but also research and community engagement.

Higher education is well placed – and in some cases uniquely suited – to addressing these characteristics. It covers the full range of human knowledge and can engage with climate change in its scientific and technological aspects, as well as its human ones – exploiting potentially the interactions between the different disciplinary areas. As a site for in-depth and sustained research, reflection and scholarship, it is ideally adapted to addressing complexity. The protection of academic freedom and the valuing of critique and deliberation lend themselves well to supporting the contested nature of climate change. While many of higher education's impacts may only become evident in the long term there are immediate changes and interventions that can be made. And finally, higher education is both international in outlook, but also global in dispersal, allowing for cross-border dialogue and interactions.

Yet these highly conducive conditions are not always realised in practice. Universities and higher education systems suffer from silo working, unhelpful hierarchies and incentives, and lack of funds. They engage in competition rather than cooperation, and often reinforce a 'monoculture of the mind' (Shiva 1993), thereby debilitating their engagement with this complex, contested, anthropogenic and global issue. A transformation is, therefore, needed to bring the best out of institution's qualities.

The early chapters of this book laid out a framework for understanding the anatomy of the university in which this transformation takes place. The framework traced the movement of influence from the five modalities of university practice – education, knowledge production, services, public debate and campus operations – through bridging actors of graduates, communities and organisations, and on to society and the ecosystem. These influences vary in terms of their intensity and reach, and operate across varying timescales, so universities need to be aware of both the immediate and tangible, but also the subtle and long-term impacts that they have on the climate crisis. Furthermore, these impacts can be either positive or negative, so the first task facing universities is to revert the historical complicity with fossil fuel industrialisation and exploitation of the natural and human worlds, and move towards sustainability and regeneration.

HEIs are inserted in national and global systems as well, so awareness is also needed of the dynamics of policy, funding and discursive trends influencing practice. Strategies of internationalisation have received bad press because of the significant carbon emissions from student flights, yet they present important opportunities in terms of global breakthroughs and awareness-raising. Institutions are inserted in a global system characterised by hierarchies and inequalities but also the potential for international communication, collaboration and mobilisation.

On the basis of the conceptual frameworks and landscape mapping of Chapters 1–5, Chapters 6–9 have provided arguments for the pathways that HEIs should take towards transformative climate action. Chapter 6 put forward the hidden benefits of engaging with climate change in teaching, through enriching the ontological, epistemological and axiological engagement of students. A broader reflection on curriculum is provided in Chapter 7, leading to a proposal for a 'topography' of climate learning, a curated learning environment designed to maximise the possibilities not only of formal teaching but also of peer collaboration and self-directed learning across the campus and beyond. The next two chapters explored possibilities of engagement with climate within and beyond the university, through research, campus operations and various forms of community engagement.

Drawing on this analytical framework, and distilling key points from the normative sections, this final chapter will outline five principles of the higher education transformation that must take place, serving as orientations that can underpin the actions to be taken.

## 1. The forge of ideas

The core business of universities is the creation, interpretation, application and distribution of ideas, concepts and theories. There are some areas to which universities can contribute but in which they are far from being the only – and perhaps they are not even the most appropriate – provider (for example, project management, practical implementation). Yet when it comes to ideas, there is no other institution in contemporary society so well suited. For all the limitations and external constraints outlined at various points during this volume, they are still a space of relative freedom and highly conducive conditions for sustained analysis, reflection, deliberation and imagining of the universe and ourselves as human beings.

In an earlier paper (McCowan 2023), a distinction was drawn between three roles of the university in relation to sustainability: the projective, the expressive and the constructive. The projective concerns itself with impact and is described by the framework put forward in Chapter 4. It projects its influence outside of itself in space and time, creating skilled professionals, new products and more effective policies. The expressive, on the other hand, moves attention to the here and now, the way the university expresses its sustainability principles through its own operations and community. These ideas were discussed in Chapter 8 through the idea of embodiment, and the need for transition to low-carbon campuses. The third role – constructive – concerns the role of the university in scrutinising, reframing and communicating the meaning of sustainable development itself.

Universities, therefore, have a vital constructive role in relation to climate change, forging the ideas that form our understandings and responses. Climate and climate change may be more concrete than the notoriously ambiguous concept of sustainable development. But deep and sustained conceptual engagement is still crucial. There is a need to constantly reassess the nature of the climate system, the nature of human societies and the interaction between them. Most obviously, the future of our societies and their relationship with the planet on which we live needs constant reassessment and reimagining. Importantly, this task is not only descriptive or analytical but also normative, in deliberating on what kind of society we want and what kind of people we want to be.

## 2. Small is beautiful

Anna Tsing (2015) in her anthropological study of the matsutake mushroom draws attention to the constant emphasis in contemporary society on 'scaling up'. The small and the local only have worth insofar as they can provide a first step, a blueprint to becoming something bigger than themselves, and they must be designed for precision scalability. But what if some things cannot be scaled up, or will see their worth diminished in doing so?

There are two ways in which the small should be valued. In order to live within the carrying capacity of the ecosystem, humanity needs to be more frugal and look to the local to a greater extent, shunning the wasteful global consumption that the privileged have enjoyed over the past century. Small-scale agriculture and production with local supply chains can ensure not only environmental protection but also worthwhile livelihoods.

Yet the 'small' that concerns us here is not only that of the sustainable lifestyles that we will ultimately have to adopt but of the transformation needed in higher education. Without a doubt, a major shift is needed across all institutions, nationally and globally. But it is a shift that can be made up of many small interventions. Scalability always pushes us to think of the largest number of people an initiative can reach, the systemwide effects. Yet, as seen in the pathways to climate impact framework, we must think about intensity as well as reach. The profound transformation of one person in the university classroom, inspired by their lecturer, may lead to a lifetime of climate activism. A group of students working with community members to protect a local water supply may lead to a transformation of relations and the emergence of innovative practices for regeneration.

It is not only that depth is important as well as breadth but that there is value in variegation. A flowering of small-scale initiatives also maximises the possibility of experimentation, diversity and innovation, allowing for cross-fertilisation and the emergence of contextualised practice. This variegation is particularly important in relation to the complex and unpredictable systems of both climate and human societies, and particularly when it comes to learning and transformation of values, given the heterogeneity of cultures and individual human beings.

None of this is to say that we should necessarily avoid the big. However vacuous, hypocritical and unlikely to be enforced they may be, international agreements on carbon emissions are also needed. Global alliances, agreements and policies on the integration of climate action into higher education are all part of the story. The point is not small instead of big but that the small is valued as it should be – not as a stepping stone, a preparation, but as a meaningful site for intense transformation. In Chapter 8, this idea was discussed in relation to prefiguration, the bringing of ultimate aims into the here and now. Prefiguring has some instrumental benefits, acting as a space for learning, experimentation and as an exemplar to others, but most importantly it is a good in itself.

## 3. Creative unruliness

Closely linked to the above principle of valuing the small is that we should not be afraid of a lack of symmetry, order and regularity in our institutions. As emphasised throughout this book, connections and synergies are undoubtedly important, in fact crucial. Too few links have been made between the different functions of the university – the courses run for its students, the research projects of its staff, the management of its estates and the work carried out for external agencies. Insights from each of these activities can be taken to the others, and in conjunction they can be mutually reinforcing. Given the multisectoral and complex nature of climate change, this transmodality – in addition to interdisciplinarity – is vital.

The usual reaction to this need for connection is to argue for unity, coordination and alignment. Yet this move is not always necessary or warranted. An example given in anarchist circles is of the international postal service, which (although with less demand in the internet era) works seamlessly between jurisdictions in spite of the absence of a global service.[1] Synergies and cross-fertilisation can be gained through networking, even in the absence of central planning. In the case of the university, light-touch central planning is particularly important since academic freedom (both within research and teaching) are vital for the success of the whole venture.

Student agency is a crucial component of universities' response to the climate crisis. Students need agency within their learning (as supported in the topography framework), yet they are not only learners but also actors. Many higher education students are already knowledgeable in and strongly committed to climate action, and their active role in campaigning and regeneration should be encouraged, both as participants and as initiators. Academic and professional staff should also be supported in developing grassroot initiatives. The resulting diversity of activity will be fruitful, but, to a large extent, it is also inevitable, given the impossibility of unifying values and actions within a university setting.

As discussed in Chapter 8, leadership is no less important in a decentralised and creatively unruly institution. Unevenness can, in fact, be curated, as is the case in the curriculum topography outlined in Chapter 7. The metaphor of the landscape garden (as opposed to the formal, geometrical one) has been used in the book to describe how institutions can be active in encouraging structures and activities that may lack standardisation but have their own organic beauty and richness.

## 4. Reimagining institutions

The previous three principles have, for the most part, aimed at reforming existing institutions. The world now has some 90,000 HEIs (MacGregor 2022) and they are not going to be replaced overnight. So, our first priority is to work with and to transform what is already there. But while working within the system we also need to work outside of it, in order to experiment and create a vision of how institutions might be different – creating *subversities* as well as *pluriversities* in Santos's (2017) terms.

A number of these experiments are already taking place. UNITIERRA (Universidad de la Tierra, or University of the Land) in Mexico, for example, subverts the whole notion of the university by having no entrance requirements, no lecturers and no curriculum. Students pursue their learning journeys in collectives, through peer-to-peer learning and apprenticeships or mentoring relationships with experts in the community, including a strong engagement with Indigenous communities and their knowledge traditions. Graduates leave with no diploma, only the learning they have actually acquired. Many of these initiatives are supported by the Ecoversities Alliance, which brings together 200 transformative learning spaces from around the world (Mandel et al. 2022).

The challenges that these institutions make to our assumptions about what a university 'looks like' – how people are admitted, how they learn, how knowledge is produced – and its very purpose are crucial to invigorating our idea of the institution. The university as institution has been extraordinarily resilient, through its near millennium of existence, and has both retained its distinctive characteristics and adapted with the changing eras. Yet like all institutions it has a tendency to fossilisation, with practices losing their connection to their original purpose, as well as usurpation by elites for maintenance of positional advantage (McCowan and Dietz 2021; McCowan 2021b). Creating space for new forms of institution at the margins of the mainstream system, possibly bearing little resemblance to our contemporary universities, is a good in itself, but these fringe universities can also come into interaction with mainstream ones for mutual benefit and enrichment of our imaginations.

## 5. University as protagonist

It is commonly heard that universities must keep abreast of new developments in technology, economics or culture. From globalisation to social media, AI and new public management, higher education has hurried to adapt, stay relevant and survive. While this openness to outside currents is important, and to some degree inevitable, the downside has been to undermine the sense of protagonism of the university, making it overly responsive and timid.

In fact, there have been various reasons for the shift towards passivity and external influence in the university. First is the movement in many systems from public funding (or stable private patronage) towards a commercial income-generating model, dependent on external markets. The lack of core funding has led to a concrete decrease in the university's ability to chart its own course, instead servicing the purchasing power of prospective students seeking employment options and commercial funders seeking knowledge to enhance their profitability. Second is the loss of confidence in its own role in the project of knowledge, in Santos's (2004) terms its crisis of hegemony and legitimacy, as the sole or most authoritative arbiter of what knowledge is valid and valuable. The epistemic unsettling of the twentieth century through poststructuralism, revival of religious fundamentalism and populism (see Chapter 3) has dented the university's sense of self, as captured in Bill Readings' (1996) *The University in Ruins*. The internet age has intensified this loss of confidence, making obsolete the role of the university as a literal store of knowledge, and empowering people to seek knowledge for themselves.

Finally, and in a somewhat contradictory manner, there is a sense that the university has been overly confident to the point of arrogance, and needs to be more humble in its dealings with the outside society.

This need for 'epistemic humility' connects strongly with the discussions in Chapter 9 about co-production and an ecology of knowledges, with the university moving towards a respect for and engagement with the knowledge forms of other communities.

This final point is indeed important, and there is a great danger in an arrogant and condescending posture taken by the university in assuming that other sectors of society and non-university populations do not hold valuable knowledge. Yet the implication here is not that university should not act, but that it should act in a different way. Despite the understandable sense of disempowerment that comes with a loss of financial autonomy and epistemic pre-eminence, none of these factors should deter universities from taking a leading role in relation to climate change and other areas of action.

Universities, in fact, do shape society – through the pathways outlined in Chapter 4, including the work and lives of graduates, and the ideas and technologies produced. But they should also embrace the sense that they *should* shape society, that instead of merely drifting with the tide, they take responsibility to improve the lives of all and the health of the ecosystem.

*

These five principles have addressed different dimensions of the work of the university. The first (forging ideas) concerns the core of the university's practice – the creation, development and dissemination of ideas. The next three (small is beautiful, creative unruliness and reimagining institutions) have addressed the way that this practice is organised. And the final one (protagonism) concerns its relationship with society.

Taken together then, these five principles allow for internal creativity and unpredictability, and yet external ambition and influence. There is something of a paradox here. If the outcomes of university practice cannot be predefined, controlled and predicted, then how can we be confident that the institution should assert itself in the society outside? For sure, such a tension has undermined confidence in the institution on the part of paymasters who would sooner be reckoning with a controllable and predictable outfit. Yet even if we cannot be sure of the outcome, we can retain faith in the process. What is constant in the university is its adherence to the quest for human understanding through open-ended enquiry (*pace* Collini 2012), combined with (it is to be hoped, although not guaranteed) a commitment to social and environmental justice on the part of those inside it. In fact, such a faith is more in keeping with the nature of climate change, given its complex and unpredictable nature:

we are better off with a process that allows for constant adaptation to changing conditions than a fixed set of outcomes that will become rapidly redundant.

There is something approaching a consensus in the literature and professional debates on higher education, sustainability and climate change that what is most needed is alignment, coordination and joined-up thinking. Silo working is seen as the greatest enemy of effective university climate action, and what we need is to pull together along a common cause across the diverse disciplines and functions of the university. This book takes a different tack. While recognising the importance of connection – and arguing strongly for its inclusion in teaching, research and community engagement through Chapters 6–9 – the book instead posits that the nature of the university makes impossible and undesirable this kind of unity. Instead, we should aim to nurture bottom-up creativity, maximising the diversity of actors, modalities and foci of climate action.

So, if variegation is what we are aiming for, is that not what we already have and therefore we need do nothing more? Certainly we have unevenness but its creative potential is far from being fulfilled. As explored in Chapter 5, global trends in higher education have promoted the roll-out of low-quality provision for commercial gain, fostered elite competition bolstered by the international rankings and their emphasis on high-level research and publications, and militated against research and community engagement in the public good. The global system has also entrenched hierarchies within and between countries, leading not to a rich diversity of mission types but to stratification of resources, quality and public recognition. These macrolevel trends have filtered down into institutions, leaving little space for academic staff to pursue either their creative curiosity-driven scholarship or social justice commitments. Students in turn in many instances have to settle for a threadbare high school type of provision, with little opportunity for the profound professional, civic and personal development needed for the transition to a sustainable future.

Creative variegation does not emerge from neglect, but in the context of a well-resourced and well-curated system. The key point is that, within these conditions, students and staff can operate in freedom to imagine, create and associate in organic, generative and novel ways. This mission diversity corresponds to the necessary agency of the university setting populated by autonomous adults, and therefore has an intrinsic justification, but is also instrumentally needed as the most effective way to respond to the complexity of the climate crisis.

The variegation operates within but also across institutions, and even internationally. Diversity and differentiation are crucial in maximising alternatives and maintaining a global ecology of knowledges, as long as they keep their horizontal nature and do not slip into vertical stratification. While universities can have global impact, 'star' institutions cannot save the planet on their own: the response to climate change will only be effective if quality HEIs are in place in all parts of the world, including low-income contexts in the Global South.

In higher education, as in all walks of life, it is worthwhile from time to time to stop and ask ourselves the question – why on earth are we doing this? What is the point of endlessly churning through graduates in various professional areas, competing for research funds and producing more and more journal articles? When it comes to knowledge and understanding, there is certainly a sense in which these are goods in themselves, and perhaps the pursuit of them is in our very nature. Yet understanding the world is also a fundamental responsibility to ourselves and to others faced with cataclysmic environmental and social challenges. Our survival and flourishing on this planet are profoundly threatened and all sectors of society are needed to bring the seismic change in our way of life – one that will be more enduring and less painful if we choose to take action now. The way out of the tunnel is only partly clear, and the terrain is constantly shifting. Societies will need their universities if they are to find a way through.

## Notes

1 The Universal Postal Union, a United Nations agency, coordinates between the postal services of different countries, rather than acting as a global postal service itself.

# References


Alexander, B. (2023). *Universities on Fire: Higher education in the climate crisis*. Baltimore: JHU Press.

Alexander, R. (2010). *Education For All, The Quality Imperative and the Problem of Pedagogy. CREATE Pathways to Access.* Research Monograph No. 20. Falmer: University of Sussex.

Almassi, B. (2012). Climate change, epistemic trust, and expert trustworthiness. *Ethics & the Environment*, 17(2), 29–49.

Altbach, P. G., Reisberg, L. and Rumbley, L. E. (2009). *Trends in Global Higher Education: Tracking an academic revolution.* A report prepared for the UNESCO 2009 World Conference on Higher Education. Paris: UNESCO.

Alves, F. et al. (2020). Climate change policies and agendas: Facing implementation challenges and guiding responses. *Environmental Science & Policy*, 104, 190–8.

Amos, R. and Carvalho, P. (2020). Locating a course on environmental justice in theories of environmental education and global citizenship. *Journal of Education for Sustainable Development*, 14(2), 140–55.

Anderson, A. (2012). Climate change education for mitigation and adaptation. *Journal of Education for Sustainable Development*, 6(2), 191–206.

Anderson, K. (2012). Climate change going beyond dangerous—Brutal numbers and tenuous hope. *Development Dialogue*, 61, 16–40. https://www.xrhull.org.uk/wp-content/uploads/Climate-Change-beyond-dangerous-Kevin-Anderson.pdf. Accessed 5 November 2024.

Arum, R. and Roksa, J. (2011). *Academically Adrift: Limited learning on college campuses*. Chicago, IL: University of Chicago Press.

Asherman, F. et al. (2016). Designing and demonstrating a master student project to explore carbon dioxide capture technology. *Journal of Chemical Education*, 93(4), 633–8.

Ashwin, P. et al. (2020). *Reflective Teaching in Higher Education*. London: Bloomsbury.

Assié-Lumumba, N. D. T. (2017). The Ubuntu paradigm and comparative and international education: Epistemological challenges and opportunities in our field. *Comparative Education Review*, 61(1), 1–21.

Atwood, M. (2003). *Oryx and Crake*. Toronto: McClelland and Stewart.

Bacevic, J. (2019). With or without U? Assemblage theory and (de)territorialising the university. *Globalisation, Societies and Education*, 17(1), 78–91.

Bangay, C. and Blum, N. (2010). Education responses to climate change and quality: Two parts of the same agenda? *International Journal of Educational Development*, 30(4), 359–68.

Barnett, R. (1997). *Higher Education: A critical business*. Buckingham: Society for Research into Higher Education and the Open University Press.

Barth, M., Godemann, J., Rieckmann, M. and Stoltenberg, U. (2007). Developing key competencies for sustainable development in higher education. *International Journal of Sustainability in Higher Education*, 8(4), 416–30.

Bell, A. (2021). *Our Biggest Experiment: An epic history of the climate crisis.* Berkeley: Counterpoint.

Berg, M. and Seeber, B. K. (2016). *The Slow Professor: Challenging the culture of speed in the academy*. Toronto: University of Toronto Press.

Bernasconi, A. (2007). Is there a Latin American model of the university? *Comparative Education Review*, 52(1), 27–52.

Berners-Lee, M. (2019). *There Is No Planet B: A handbook for the make or break years*. Cambridge: Cambridge University Press.

Bernstein, B. (1971). On the classification and framing of educational knowledge. In M. F. D. Young (ed.) *Knowledge and Control*. London: Collier-Macmillan.

Biggs, J. (1999). What the student does: Teaching for enhanced learning. *Higher Education Research & Development*, 18(1), 57–75.

Binagwaho, A. et al. (2022). *Knowledge-driven Actions: Transforming higher education for global sustainability. Independent expert group on the universities and the 2030 Agenda*. Paris: UNESCO.

Bjørkdahl, K., Duharte, F. and Santiago, A. (eds) (2022). *Academic Flying and the Means of Communication*. London: Palgrave Macmillan.

Bloom, H. (2014). *The Western Canon: The books and school of the ages*. Boston: Houghton Mifflin Harcourt.

Blumenstyk, G. (2014). *American Higher Education in Crisis? What everyone needs to know.* New York: Oxford University Press.

Boggs, C. (1978). Marxism, prefigurative communism, and the problem of workers' control. *Radical America*, 11/12(1), 99–122.

Booker, C. (2009). *The Real Global Warming Disaster: Is the obsession with 'climate change' turning out to be the most costly scientific blunder in history?* London: Continuum.

Booth, A., Earley, S., Aben, K., Otter, B., Corrigal, T. and Ray, C. (2020). Action learning partnerships: carbon, commerce and community co-learning at a Canadian university. *International Journal of Sustainability in Higher Education*, 21(5), 943–57.

Boslough, M. (2021). A critical review of Steven Koonin's 'Unsettled'. *Yale Climate Connections*, 25 May. Available at: https://yaleclimateconnections.org/2021/05/a-critical-review-of-steven-koonins-unsettled/. Accessed 22 May 2024.

Bowers, C. A. and Apffel-Marglin, F. (eds) (2005). *Re-thinking Freire: Globalization and the environmental crisis*. London: Routledge.

Brown, E. and McCowan, T. (2018). Buen vivir: reimagining education and shifting paradigms (Forum). *Compare*, 48(2), 317–23.

Browne, M. N. and Freeman, K. (2000). Distinguishing features of critical thinking classrooms. *Teaching in Higher Education*, 5(3), 301–9.

Brugger, J. and Crimmins, M. (2015). Designing institutions to support local-level climate change adaptation: Insights from a case study of the US Cooperative Extension System. *Weather, Climate, and Society*, 7(1), 18–38.

Bryan, A. (2020). Affective pedagogies: Foregrounding emotion in climate change education. *Policy and Practice: A Development Education Review*, 30, 8–30.

Burandt, S. and Barth, M. (2010). Learning settings to face climate change. *Journal of Cleaner Production*, 18(7), 659–65.

Burgen, S. (2022). Barcelona students to take mandatory climate crisis module from 2024. *The Guardian*, 12 November.

Bush, D., Sieber, R., Seiler, G. and Chandler, M. (2017). University-level teaching of Anthropogenic Global Climate Change (AGCC) via student inquiry, *Studies in Science Education*, 53(2), 113–36.

Bussey, M. (2010). Fostering new values and stories in higher education: What climate change brings to the educational table. *Journal of Future Studies*, 15(1), 101–8.

Bynner, J., Dolton, P., Feinstein, L., Makepiece, G., Malmberg, L. and Woods, L. (2003). *Revisiting the benefits of higher education: A report by the Bedford Group for Lifecourse and Statistical Studies, Institute of Education*. Bristol: Higher Education Funding Council for England.

Byskov, M. F. and Hyams, K. (2022). Epistemic injustice in climate adaptation. *Ethical theory and moral practice*, 25(4), 613–34.

Campbell, A. C., Nguyen, T. and Stewart, M. (2023). Promoting international student mobility for sustainability? Navigating conflicting realities and emotions of international educators. *Journal of Studies in International Education*, 27(4), 621–37.

CANIE (2022). The CANIE Accord. Available at: https://canie.org/the-canie-accord. Accessed 27 December 2022.

Capra, F. (1996). *The Web of Life: A new synthesis of mind and matter*. London: Flamingo.

Carayannis, E. G. and Campbell, D. F. (2009). 'Mode 3' and 'Quadruple Helix': toward a 21st century fractal innovation ecosystem. *International Journal of Technology Management*, 46(3–4), 201–34.

Carayannis, E. G., Barth, T. D. and Campbell, D. F. (2012). The Quintuple Helix innovation model: global warming as a challenge and driver for innovation. *Journal of Innovation and Entrepreneurship*, 1, 1–12.

Carpentier, V. (2012). Public-private substitution in higher education: Has cost-sharing gone too far? *Higher Education Quarterly*, 66, 363–90.

Carpentier, V. (2019). The historical expansion of higher education in Europe: Spaces, shapes and rationales. In J. L. Rury and E. H. Tamura (eds), *The Oxford Handbook of the History of Education*. Oxford: Oxford University Press.

Carson, R. (2022 [1962]) *Silent Spring*. London: Penguin.

Chambers, R. (1994). The origins and practice of participatory rural appraisal. *World Development*, 22(7), 953–69.

Clark, B. R. (1983). *The Higher Education System: Academic organization in cross-national perspective*. Berkeley: University of California Press.

Climate-U (2023). *Five Conditions for Participatory Action Research to Enhance Universities' Contributions to Climate Justice*. Transforming Universities for a Changing Climate, Working Paper Series No. 19.

Coady, D. and Corry, R. (2013). *The Climate Change Debate: An epistemic and ethical enquiry*. London: Palgrave Macmillan.

Collini, S. (2012). *What Are Universities For?* London: Penguin.

Cook, J. et al. (2013). Quantifying the consensus on anthropogenic global warming in the scientific literature. *Environmental Research Letters*, 8(2), 024024.

Cook, J., et al. (2016). Consensus on consensus: A synthesis of consensus estimates on human-caused global warming. *Environmental Research Letters*, 11(4), 048002.

Cordero, E. C., Centeno, D. and Todd, A. M. (2020). The role of climate change education on individual lifetime carbon emissions. *PloS one*, 15(2), e0206266.

Cotton, D., Bailey, J. and Tosdevin, M. (2020). Higher education and the climate emergency: Exploring the hidden curriculum of the campus. In T. Hinchcliffe (ed.) *The Hidden Curriculum of Higher Education* (Advance HE). Available at: https://s3.eu-west-2.amazonaws.com/assets.creode.advancehe-document-manager/documents/advance-he/AdvHE_Hidden%20Curriculum_1597673443.pdf. Accessed 30 May 2024.

Crenshaw, K. W. (1991). Mapping the margins: Intersectionality, identity politics, and violence against women of color. *Stanford Law Review*, 43(6), 1241–99.

Crow, M. M. (2010). Organizing teaching and research to address the grand challenges of sustainable development. *BioScience*, 60(7), July/August, 488–9.

Czerniewicz, L., Goodier, S. and Morrell, R. (2017). Southern knowledge online? Climate change research discoverability and communication practices. *Information, Communication & Society*, 20(3), 386–405.

Davies, J. C. and Dunk, R. M. (2015). Flying along the supply chain: Accounting for emissions from student air travel in the higher education sector. *Carbon Management*, 6(5–6), 233–46.

De Wit, H. and Altbach, P. (2020). Time to cut international education's carbon footprint. *University World News*, 11 January. https://www.universityworldnews.com/post.php?story=20200108084344396. Accessed 4 November 2024.

De Wit, H. and Altbach, P. (2021). Internationalization in higher education: Global trends and recommendations for its future. *Policy Reviews in Higher Education*, 5(1), 28–46.

Del Monte, P. and Posholi, L. (2021). Decolonial perspectives on education and international development. In T. McCowan and E. Unterhalter (eds), *Education and International Development: An Introduction*. London: Bloomsbury.

Delingpole, J. (2012). *Watermelons: How Environmentalists are Killing the Planet, Destroying the Economy and Stealing Your Children's Future*. London: Biteback.

Dworkin, R. (1996). We need a new interpretation of academic freedom. *Academe*, 82(3), 10–15.

Ellis, H. (2019). Beyond the university: Higher education institutions across time and space. In T. Fitzgerald (ed.) *Handbook of Historical Studies in Education: Debates, tensions, and directions*, 1–17. Springer International Handbooks of Education. Singapore: Springer.

Epstein, B. (1988). *Politics of Prefigurative Community: The Nonviolent Direction Action Movement*. In M. Davis and M. Sprinker (eds) *Reshaping the US Left: Popular struggles in the 1980s*, 63–92. London: Verso.

Etzkowitz, H. and Leydesdorff, L. (1995). The Triple Helix—University-industry-government relations: A laboratory for knowledge based economic development. *EASST review*, 14(1), 14–19. Available at: https://ssrn.com/abstract=2480085. Accessed 8 November 2024.

Evans, J. and Karvonen, A. (2014). 'Give me a laboratory and I will lower your carbon footprint!'—Urban laboratories and the governance of low-carbon futures. *International Journal of Urban and Regional Research*, 38(2), 413–30.

Facer, K. (2020). *Beyond Business as Usual: Higher Education in the Era of Climate Change*. Oxford: Higher Education Policy Institute.

Facer, K. (2023). Possibility and the temporal imagination. *Possibility Studies & Society*, 1(1–2), 60–6.

Facer, K., Lotz-Sisitka, H., Ogbuigwe, A., Vogel, C. and Barrineau, S. (2020). *Climate Change and Education*. Transforming Education for Sustainable Futures (TESF) Briefing Paper. Bristol: TESF.

Fahey, S. J. (2012). Curriculum change and climate change: Inside outside pressures in higher education, *Journal of Curriculum Studies*, 44(5), 703–22.

Farnell, T., Benneworth, P., Ćulum Ilić, B., Seeber, M., Šćukanec Schmidt, N. (2020). *TEFCE Toolbox: An institutional self-reflection framework for community engagement in higher education*. Zagreb: Institute for the Development of Education.

Fillmore, H. M., Singletary, L. and Phillips, J. (2018). Assessing tribal college priorities for enhancing climate adaptation on reservation lands. *Journal of Contemporary Water Research & Education*, 163(1), 64–78.

Findler, F., Schönherr, N., Lozano, R., Reider, D. and Martinuzzi, A. (2019a) The impacts of higher education institutions on sustainable development. A review and conceptualization. *International Journal of Sustainability in Higher Education*, 20(1), 23–38.

Findler, F., Schönherr, N., Lozano, R. and Stacherl, B. (2019b) Assessing the impacts of higher education institutions on sustainable development—an analysis of tools and indicators. *Sustainability*, 11(1), 59.

Ford, J. D., Cameron, L., Rubis, J., Maillet, M., Nakashima, D., Cunsolo, A. and Pearce, T. (2016). Including indigenous knowledge and experience in IPCC assessment reports. *Nature Climate Change*, 6(4), 349–53.

Frediani, A. and Nussey, C. (2021). *A Protocol for Participatory Action Research into Climate Justice: Principles and tools*. Transforming Universities for a Changing Climate, Working Paper Series No. 3.

Freire, P. (1970). *Pedagogy of the Oppressed*. London: Penguin.

Freire, P. (1994). *Pedagogy of Hope: Reliving Pedagogy of the Oppressed*. New York: Continuum.

Fricker, M. (2007). *Epistemic Injustice: Power and the Ethics of Knowing*. Oxford: Oxford University Press.

Fry, H., Ketteridge, S. and Marshall, S. (eds) (2008). *A Handbook for Teaching and Learning in Higher Education: Enhancing academic practice*. London: Routledge.

Fung, D. (2017). *A Connected Curriculum for Higher Education*. London: UCL Press.

Fyfe, A., Coate, K., Curry, S., Lawson, S., Moxham, N. and Røstvik, C. M. (2017). Untangling Academic Publishing: A history of the relationship between commercial interests, academic prestige and the circulation of research. Discussion Paper. University of St Andrews.

Gadotti, M. (2000). *Pedagogia da terra: Ecopedagogia e educação sustentável.* Buenos Aires: CLACSO.

García del Amo, D., Mortyn, P.G. and Reyes García, V. (2020). Including indigenous and local knowledge in climate research: An assessment of the opinion of Spanish climate change researchers, *Climatic Change*, 160(1), 67–88.

Gates, B. (2021). *How To Avoid a Climate Disaster: The solutions we have and the breakthroughs we need*. London: Allen Lane.

Gayle, D. (2023). Students occupy schools and universities across Europe in climate protest. *The Guardian*, 5 May.

Gazni, A., Sugimoto, C. R. and Didegah, F. (2012). Mapping world scientific collaboration: Authors, institutions, and countries. *Journal of the American Society for Information Science and Technology*, 63(2), 323–35.

Ghosh, A. (2016). *The Great Derangement: Climate change and the unthinkable*. Chicago: University of Chicago Press.

Gill, M. (2021). High flying business schools: Working together to address the impact of management education and research on climate change. *Journal of Management Studies*, 58(2), 554–61.

Giroux, H. A. and Penna, A. N. (1979). Social education in the classroom: The dynamics of the hidden curriculum. *Theory & Research in Social Education*, 7(1), 21–42.

Gladwell, M. (2000). *The Tipping Point: How little things can make a big difference*. London: Little, Brown.

Goldman, M. J., Turner, M. D. and Daly, M. (2018). A critical political ecology of human dimensions of climate change: Epistemology, ontology, and ethics. *Wiley Interdisciplinary Reviews: Climate Change*, 9(4), e526.

Goldrick-Rab, S., (2016). *Paying the Price: College costs, financial aid, and the betrayal of the American dream*. Chicago, IL: University of Chicago Press.

Gomes, L. A. (2020). Ambientalização Curricular nos Cursos de Licenciatura na Universidade Federal do Oeste do Pará. Unpublished PhD thesis, Federal University of the West of Pará.

Grady-Benson, J. and Sarathy, B. (2016). Fossil fuel divestment in US higher education: Student-led organising for climate justice. *Local Environment*, 21(6), 661–81.

Greer, K., Sheldrake, R., Rushton, E., Kitson, A., Hargreaves, E. and Walshe, N. (2023). *Teaching Climate Change and Sustainability: A survey of teachers in England*. London: University College London.

Grieve, T. and Mitchell, R. (2020). Promoting meaningful and equitable relationships? Exploring the UK's Global Challenges Research Fund (GCRF) funding criteria from the perspectives of African partners. *The European Journal of Development Research*, 32(3), 514–28.

Gutmann, A. (1987). *Democratic Education*. Princeton: Princeton University Press.

Haddock-Fraser, J., Rands, P. and Scoffham, S. (2018). *Leadership for Sustainability in Higher Education*. London: Bloomsbury Academic.

Hage, S. A. M., Pereira, E. A. D, Barros, O. F., Lima, M. V. C., Castro, W. S., Taveres, G. C., Ribeiro, C. A. A., Araújo, M. N. C. (2023). *Territorial Governance Collectives: Instrument of formation, production of knowledge and intervention for the realization of climate justice in the Amazon*. Transforming Universities for a Changing Climate, Working Paper Series No. 16.

Hansen, J. A. and Lehmann, M. (2006). Agents of change: Universities as development hubs. *Journal of Cleaner Production*, 14(9–11), 820–9.

Haskins, L. and Prideaux, F. (2023). Just Stop Oil paints Oxford University orange. *The Oxford Blue*, 10 October.

Hawkey, K. (2023). *History and the Climate Crisis: Environmental history in the classroom*. London: UCL Press.

Haydon, G. (1977). The right to education and compulsory schooling, *Educational Philosophy and Theory*, 9(1), 1–15.

Hazelkorn, E. (2011). *Rankings and the Reshaping of Higher Education: The battle for world class excellence*. Basingstoke: Palgrave Macmillan.

Head, B. W. and Alford, J. (2015). Wicked problems: Implications for public policy and management. *Administration & Society*, 47(6), 711–39.

Helm, D. (2020). *Net Zero: How we stop causing climate change*. London: HarperCollins.

Helmers, E., Chang, C. C. and Dauwels, J. (2021). Carbon footprinting of universities worldwide: Part I—objective comparison by standardized metrics. *Environmental Sciences Europe*, 33, 1–25.

Henderson, J., Bieler, A. and McKenzie, M. (2017). Climate change and the Canadian higher education system: An institutional policy analysis. *Canadian Journal of Higher Education*, 47(1), 1–26.

Hentschke, G. C., Lechuga, V. M. and Tierney, W. G. (eds). (2023). *For-profit Colleges and Universities: Their markets, regulation, performance, and place in higher education*. New York: Routledge.

HESA (n.d.). Definitions: Estate management. Available at: https://www.hesa.ac.uk/support/definitions/estates. Accessed 8 November 2024.

Hess, D. J. and Collins, B. M. (2018). Climate change and higher education: Assessing factors that affect curriculum requirements. *Journal of Cleaner Production*, 170, 1451–8.

Hickel, J. (2021). *Less is More: How degrowth will save the world*. London: Penguin.

Hickman, C. et al. (2021). Climate anxiety in children and young people and their beliefs about government responses to climate change: A global survey. *The Lancet Planetary Health*, 5(12), e863–e873.

Hillmer-Pegram, K. C., Howe, P. D., Greenberg, H. and Yarnal, B. (2012). A geographic approach to facilitating local climate governance: From emissions inventories to mitigation planning. *Applied Geography*, 34, 76–85.

Hinchcliffe, T. (ed.) (2020). *The Hidden Curriculum of Higher Education*. York: Advance HE. Available at: https://s3.eu-west-2.amazonaws.com/assets.creode.advancehe-document-manager/documents/advance-he/AdvHE_Hidden%20Curriculum_1597673443.pdf. Accessed 30 May 2024.

Hoare, A. and Johnston, R. (2011). Widening participation through admissions policy – a British case study of school and university performance. *Studies in Higher Education*, 36(1), 21–41.

Holloway, A. and Fortune, G. (2018). *Beyond fragility: Advancing skilled human capital for disaster risk reduction and resilience in Africa.* Background paper for UNISDR Global Assessment of Risk 2019. United Nations Office for Disaster Risk Reduction (UNISDR). https://www.unisdr.org/we/inform/publications/65971. Accessed 4 November 2024.

Ignatieff, M. (2018). *Academic Freedom and the Future of Europe*. Centre for Global Higher Education, Working Paper No. 40. https://doi.org/10.5287/ora-zrkqoenxx.

Ilieva, J. and Tsiligiris, V. (2021). Globally engaged and locally relevant: Revisiting higher education. *International Higher Education*, 106, 13–14.

Inhofe, J. M. (2012). *The Greatest Hoax: How the global warming conspiracy threatens your future*. Washington, DC: WND Books.

IPCC. (2018). *Global Warming of 1.5°C*. An IPCC Special Report on the impacts of global warming of 1.5°C above pre-industrial levels and related global greenhouse gas emission pathways, in the context of strengthening the global response to the threat of climate change, sustainable development, and efforts to eradicate poverty. Available at: https://www.ipcc.ch/sr15/. Intergovernmental Panel on Climate Change. Accessed 1 November 2018.

Jacobi, P. (2003). Educação ambiental, cidadania e sustentabilidade. *Cadernos de pesquisa*, 118, 189–206.

Jackson, P. (1968). *Life in Classrooms.* New York: Holt, Rinehart and Winston.

Jickling, B. and Wals, A. (2008). Globalization and environmental education: Looking beyond sustainable development. *Journal of Curriculum Studies*, 40(1), 1–21.

Jimenez, J. and Kabachnik, P. (2023). Indigenizing environmental sustainability curriculum and pedagogy: Confronting our global ecological crisis via Indigenous sustainabilities. *Teaching in Higher Education*, 28(5), 1095–107.

Kahn, P., Moreau, M. P. and Gagnon, J. (2024). Precarity and illusions of certainty in higher education teaching. *Teaching in Higher Education*, 29(3), 699–706.

Kahn, R. V. (2010). *Critical Pedagogy, Ecoliteracy, & Planetary Crisis: The ecopedagogy movement*. New York: Peter Lang.

Kember, D. and Gow, L. (1994). Orientations to teaching and their effect on the quality of student learning. *The Journal of Higher Education*, 65(1), 58–74.

Kerr, C. (1963). *The Uses of the University*. New York: Harper Torchbooks.

Kerr, D. (1999). 'Citizenship education in the curriculum: An international review'. *The School Field*, X(3/4), 5–32.

King, Y. (1995). Engendering a peaceful planet: Ecology, economy, and ecofeminism in contemporary context. *Women's Studies Quarterly*, 23(3/4), 15–21.

Kinol, A., Miller, E., Axtell, H., Hirschfeld, I., Leggett, S., Si, Y. and Stephens, J. C. (2023). Climate justice in higher education: A proposed paradigm shift towards a transformative role for colleges and universities. *Climatic Change*, 176(2), 15.

Kitagawa, K. (2017). Situating preparedness education within public pedagogy, *Pedagogy, Culture & Society*, 25(1), 1–13.

Klein, N. (2014). *This Changes Everything: Capitalism vs. the climate*. New York: Simon and Schuster.

Klein, N. (2019). *On Fire. The Burning Case for a Green New Deal.* New York: Simon and Schuster.

Knight, J. (2022). *Knowledge Diplomacy in International Relations and Higher Education*. Cham: Springer.

Komatsu, H., Rappleye, J. and Silova, I. (2021). Student-centered learning and sustainability: Solution or problem? *Comparative Education Review*, 65(1), 6–33.

König, A. (ed.) (2013). *Regenerative Sustainable Development of Universities and Cities: The role of living laboratories*. Cheltenham, UK: Edward Elgar.

König, A. and Evans, J. (2013). Introduction: Experimenting for sustainable development? Living laboratories, sustainable development and the role of the university. In A. König (ed.), *Regenerative Sustainable Development of Universities and Cities: The role of living laboratories*. Cheltenham, UK: Edward Elgar.

Koonin, S. (2021). *Unsettled. What Climate Science Tells Us, What It Doesn't, and Why It Matters*. Dallas: BenBella.

Kuhn, D. (1999). A developmental model of critical thinking. *Educational Researcher*, 28(2), 16–25+46. http://www.jstor.org/stable/1177186.

Kukkonen, A., Stoddart, M. C. and Ylä-Anttila, T. (2020). Actors and justifications in media debates on Arctic climate change in Finland and Canada: A network approach, *Acta Sociologica*, 64(1), 103–17.

Kwauk, C. and Casey, O. (2021). A new green learning agenda: Approaches to quality education for climate action. Washington, DC: Center for Universal Education at The Brookings Institution.

Læssøe, J., Schnack, K., Breiting, S., Rolls, S., Feinstein, N. and Goh, K. C. (2009). *Climate Change and Sustainable Development: The Response from Education.* A cross-national report from international alliance of leading education institutes. Aarhus: The Danish School of Education, Aarhus University.

Lage, M. J., Platt, G. J. and Treglia, M. (2000). Inverting the classroom: A gateway to creating an inclusive learning environment. *The Journal of Economic Education*, 31(1), 30–43.

Latter, B. and Capstick, S. (2021). Climate emergency: UK universities' declarations and their role in responding to climate change. *Frontiers in Sustainability*, 2, 660596.

Laurillard, D. (2002). *Rethinking University Teaching: A conversational framework for the effective use of learning technologies*. London: Routledge.

Laurillard, D. and Kennedy, E. (2017). *The Potential of MOOCs for Learning at Scale in the Global South*. Centre for Global Higher Education, Working Paper No. 31.

Le Roux, C. and Pretorius, M. (2016). Conceptualizing the limiting issues inhibiting sustainability embeddedness. *Sustainability*, 8(4), 364.

Leal Filho, W. (ed.). (2010). *Universities and Climate Change: Introducing climate change to university programmes*. Heidelberg: Springer.

Leal Filho, W., Morgan, E. A., Godoy, E. S., Azeiteiro, U. M., Bacelar-Nicolau, P., Veiga Ávila, L., MacLean, C. and Hugé, J. (2018). Implementing climate change research at universities: Barriers, potential and actions. *Journal of Cleaner Production*, 170, 269–77.

Lee, J. J. and Lundemo, O. A. (2021). Why sustainability is not enough in international education. *University World News*, 5 June. https://www.universityworldnews.com/post.php?story=20210531123436636. Accessed 4 November 2024.

Lehtonen, A., Salonen, A. O. and Cantell, H. (2019). Climate change education: A new approach for a world of wicked problems. In J. W. Cook (ed.) *Sustainability, Human Well-Being, and the Future of Education.* London: Palgrave Macmillan.

Leibowitz, B. and Bozalek, V. (2018). Towards a slow scholarship of teaching and learning in the South. *Teaching in Higher Education*, 23(8), 981–94.

Lenton, T. M., Rockström, J., Gaffney, O., Rahmstorf, S., Richardson, K., Steffen, W. and Schellnhuber, H. J. (2019). Climate tipping points—too risky to bet against. *Nature*, 575 (7784), 592–5.

Levin, K., Cashore, B., Bernstein, S. and Auld, G. (2012). Overcoming the tragedy of super wicked problems: Constraining our future selves to ameliorate global climate change. *Policy Sciences*, 45, 123–52.

Lewis, S. and Maslin, M. (2018). *The Human Planet: How we created the Anthropocene*. London: Pelican.

Ley, A. J. (2018). Mobilizing doubt: The legal mobilization of climate denialist groups. *Law and Policy*, 40(3), 221–42.

Lipman, M. (1988). Critical thinking: What can it be? *Educational Leadership*, 46, 38–43.

Lomborg, B. (2007). *Cool It. The Skeptical Environmentalist's Guide to Global Warming.* New York: Knopf.

Lotz-Sisitka, H., Wals, A. E., Kronlid, D. and McGarry, D. (2015). Transformative, transgressive social learning: Rethinking higher education pedagogy in times of systemic global dysfunction. *Current Opinion in Environmental Sustainability*, 16, 73–80.

Lovelock, J. (1972). Gaia as seen through the atmosphere. *Atmospheric Environment*, 6, 579–80.

Lubchenco, J. and Rapley, C. (2020). Our moment of truth: The social contract realized? *Environmental Research Letters*, 15(11), 110201.

Lynas, M., Houlton, B. Z. and Perry, S. (2021). Greater than 99% consensus on human caused climate change in the peer-reviewed scientific literature. *Environmental Research Letters*, 16(11), 114005.

MacGregor, K. (2022). Higher education report charts rise of the Global South. *University World News*, 12 March.

Macintyre, T., Lotz-Sisitka, H., Wals, A., Vogel, C. and Tassone, V. (2018). Towards transformative social learning on the path to 1.5 degrees. *Current Opinion in Environmental Sustainability*, 31, 80–7.

Mance, H. (2016). Britain has had enough of experts, says Gove. *Financial Times*, 3 June. https://www.ft.com/content/3be49734-29cb-11e6-83e4-abc22d5d108c. Accessed 5 November 2024.

Mandel, U., Lopez-Amaro, G. and Teamey, K. (2022). Ecoversities Alliance: A five-year experiment in cosmopolitical learning. *Educação & Realidade*, 46, e118644.

Mann, M. E. (2022). Throwing soup at art shifted people's views of climate protests—But maybe not in the right way. *Time*, 15 November.

Mann, M. E. and Toles, T. (2016). *The Madhouse Effect: How climate change denial is threatening our planet, destroying our politics, and driving us crazy*. New York: Columbia University Press.

Mann, M. E., Bradley, R. S., Hughes, M. K. (1998). Global-scale temperature patterns and climate forcing over the past six centuries. *Nature*, 392, 779–87.

Marginson, S. (2011). Higher education and public good. *Higher Education Quarterly*, 65(4), 411–33.

Marginson, S. (2018). Public/private in higher education: A synthesis of economic and political approaches. *Studies in Higher Education*, 43(2), 322–37.

Marginson, S. (2024). *The New Geo-politics of Higher Education 2: Between nationalism and globalism*. Centre for Global Higher Education, Working Paper No. 108.

Marginson, S. and Considine, M. (2000). *The Enterprise University: Power, governance, and reinvention in Australia*. Cambridge, UK: Cambridge University Press.

Margolis, E. (ed.) (2001). *The Hidden Curriculum in Higher Education*. New York: Routledge.

Margulis, L. (1998). *Symbiotic Planet: A new look at evolution*. Amherst: Basic Books.

Marshall, G. (2014). *Don't Even Think About It: Why our brains are wired to ignore climate change*. London: Bloomsbury.

Martin, C. (2022). *The Right to Higher Education*. Oxford: Oxford University Press.

Masson-Delmotte, V. et al. (2021). *Climate Change 2021: The Physical Science Basis*. Contribution of working group I to the sixth assessment report of the Intergovernmental Panel on Climate Change.

Mazhar, M. U., Bull, R. and Lemon, M. (2017). Critical success factors for embedding carbon management in organizations: Lessons from the UK higher education sector. *Carbon Management*, 8(5–6), 379–92.

Mazigo, A., Mwita, E., Machumu, M. and Mayenga, J. (2023). *Unlocking and Supporting Community-led Climate Actions in the Coastal Communities of Tanzania*. Transforming Universities for a Changing Climate, Working Paper Series No. 20.

McCowan, T. (2004). The growth of private higher education in Brazil: Implications for equity and quality. *Journal of Education Policy*, 19(4), 453–72.

McCowan, T. (2009). *Rethinking Citizenship Education: A Curriculum for Participatory Democracy*. London: Continuum.

McCowan, T. (2010). School democratization in prefigurative form: Two Brazilian experiences. *Education, Citizenship and Social Justice*, 5(1), 21–41.

McCowan, T. (2013). *Education as a Human Right: Principles for a Universal Entitlement to Learning*. London: Bloomsbury.

McCowan, T. (2014). *Can Higher Education Solve Africa's Job Crisis? Understanding Graduate Employability in Sub-Saharan Africa*. Policy brief. Manchester: British Council.

McCowan, T. (2015a). Theories of development. In T. McCowan and E. Unterhalter (eds), *Education and International Development: An introduction*, 31–48. London: Bloomsbury.

McCowan, T. (2015b). Should universities promote employability? *Theory and Research in Education*, 13(3), 267–85.

McCowan, T. (2016). Three dimensions of equity of access in higher education. *Compare*, 46(4), 645–65.

McCowan, T. (2017). Higher education, unbundling and the end of the university as we know it. *Oxford Review of Education*, 43(6), 733–48.

McCowan, T. (2018). Five perils of the impact agenda in higher education. *London Review of Education*, 16(2), 279–95.

McCowan, T. (2019). *Higher Education for and beyond the Sustainable Development Goals*. London: Palgrave Macmillan.

McCowan, T. (2020). *The Impact of Universities on Climate Change: A theoretical framework*. Transforming Universities for a Changing Climate, Working Paper Series No. 1.

McCowan, T. (2021a). *Climate Change in Higher Education: A curriculum topography approach*. Transforming Universities for a Changing Climate, Working Paper Series No. 6.

McCowan, T. (2021b). Deinstitutionalisation and renewal in higher education. *Educação & Realidade*, 46(4), e117607.

McCowan, T. (2022). Can we measure universities' impact on climate change? *International Higher Education*, 111, 17–19.

McCowan, T. (2023). *Tertiary Education and the Sustainability Agenda*. Centre for Global Higher Education, Working Paper No. 90.

McCowan, T. and Dietz, G. (2021). Within or beyond the university? Experiences of alternative higher education. *Educação & Realidade*, 46(4), e120754.

McCowan, T., Leal Filho, W. and Brandli, L. (eds) (2021). *Universities Facing Climate Change and Sustainability*. Hamburg: Global Universities Leaders Council.

McCowan, T., Omingo, M., Schendel, R., Adu-Yeboah, C. and Tabulawa, R. (2022). Enablers of pedagogical change within universities: Evidence from Kenya, Ghana and Botswana. *International Journal of Educational Development*, 90, 102558.

McDowell, G. (2003). Engaged universities: Lessons from the Land-Grant universities and extension. *Annals of the American Academy of Political and Social Science*, 585 (Higher Education in the Twenty-First Century), 31–50.

McMahon, W. W. (2009). *Higher Learning, Greater Good*. Baltimore: John Hopkins Press.

Merchant, C. (1980). *The Death of Nature: Women, ecology, and the scientific revolution*. New York: HarperCollins.

Meyer, H.-D., St John, E. P., Chankseliani, M. and Uribe, L. (eds) (2013). *Fairness in Access to Higher Education in a Global Perspective: Reconciling excellence, efficiency, and justice*. Rotterdam: Sense.

Mill, J. S. (1991 [1861]). *On Liberty and Other Essays*. Oxford: Oxford University Press.

Misiaszek, G. W. (2020a). *Ecopedagogy: Critical environmental teaching for planetary justice and global sustainable development*. London: Bloomsbury.

Misiaszek, G. W. (2020b). Ecopedagogy: Teaching critical literacies of 'development', 'sustainability', and 'sustainable development'. *Teaching in Higher Education*, 25(5), 615–32.

Misiaszek, G. W. and Rodrigues, C. (2022). Six critical questions for teaching justice-based environmental sustainability (JBES) in higher education, *Teaching in Higher Education*, 28(5), 903–17.

Molebatsi, P. and McCowan, T. (2022). Indicators of higher education and the public good in Africa. *Journal of Higher Education in Africa*, 20(2), 185–210.

Molthan-Hill, P., Worsfold, N., Nagy, G. J., Leal Filho, W. and Mifsud, M. (2019). Climate change education for universities: A conceptual framework from an international study. *Journal of Cleaner Production*, 226, 1092–101.

Monroe, M. C. and Oxarart, A. (2019). Integrating research and education: developing instructional materials to convey research concepts, *BioScience*, 69(4), 282–91.

Monroe, M., Plate, R., Oxarart, A., Bowers, A. and Chaves, W. (2019). Identifying effective climate change education strategies: A systematic review of the research, *Environmental Education Research*, 25(6), 791–812.

Morano, M. (2018). *The Politically Incorrect Guide to Climate Change.* Washington, DC: Regnery.

Moscovitz, H. and Sabzalieva, E. (2023). Conceptualising the new geopolitics of higher education. *Globalisation, Societies and Education*, 21(2), 149–65.

Mtawa, N. N. (2019). *Human Development and Community Engagement Through Service-Learning*. Suiza: Springer Nature.

Mutanga, O. and Marovah, T. (eds). (2024). *Southern Theories: Contemporary and future challenges*. London: Routledge.

Myers, K. F., Doran, P. T., Cook, J., Kotcher, J. E. and Myers, T. A. (2021). Consensus revisited: quantifying scientific agreement on climate change and climate expertise among Earth scientists 10 years later. *Environmental Research Letters*, 16(10), 104030.

Newman, J. (2021). USA. In T. McCowan, W. Leal Filho and L. Brandli (eds), *Universities Facing Climate Change and Sustainability*. Hamburg: Global Universities Leaders Council.

Nikula, P. T. (2019). Towards carbon neutral international education. *Forum*, EAIE Magazine, Winter, 12–13.

Norgaard, K. M. (2011). *Living in Denial: Climate change, emotions, and everyday life*. Cambridge: MIT Press.

Norões, K. and Costa, B. (2012). Affirmative policies in Brazil: Black movements and public higher education. *Educational Thought*, 9(1), 24–31.

Nugent, C. (2021). The unexpected ways climate change is reshaping college education. *Time*, 16 April. Available at: https://time.com/5953399/college-education-climate-change/. Accessed 22 July 2021.

Nussbaum, M. (1997). *Cultivating Humanity: A classical defense of reform in liberal education.* Cambridge, MA: Harvard University Press.

Nussey, C. (2021). Rethinking the unthinkable: what can educational engagements with culture offer the climate crisis? British Council Cultural Relations Collection. Available at: https://www.britishcouncil.org/research-insight/rethinking-the-unthinkable-climate-crisis. Accessed 12 October 2021.

Nussey, C., Sanchez Tyson, L., Dandare, K., Perry, J. and McCowan, T. (2023). *How Are Universities Responding to the Challenges of Climate Change? A systematic review of literature*. Transforming Universities for a Changing Climate, Working Paper Series No. 14.

Oberman, R. and Sainz, G. M. (2021). Critical thinking, critical pedagogy and climate change education. In A. M. Kavanagh, F. Waldron and B. Mallon (eds), *Teaching for Social Justice and Sustainable Development across Primary Curriculum*, 69–83. London: Routledge.

OECD. (2012). How 'green' are today's 15-year-olds? *PISA in Focus*, No. 15. Paris: OECD Publishing. https://doi.org/10.1787/5k9bdt3vsbf0-en.

Ojala, M. (2016). Facing anxiety in climate change education: From therapeutic practice to hopeful transgressive learning. *Canadian Journal of Environmental Education (CJEE)*, 21, 41–56.

Oketch, M., McCowan, T. and Schendel, R. (2014). *The Impact of Tertiary Education on Development: A rigorous literature review.* London: Department for International Development.

Oldac, Y. I., Nkansah, J. O. and Yang, L. (2023). 'West is must, the rest is optional': epistemic injustice and positional good in international research collaboration. *Higher Education*, 88(2), 1–18.

Olivera Rodríguez, I. (2017). Las potencialidades del proyecto educativo de la Universidad Veracruzana Intercultural: una crítica al desarrollo desde la noción del Buen vivir. *Revista de la Educación Superior*, 46(181), 19–35.

Oreskes, N. (2004). The scientific consensus on climate change. *Science*, 306(5702), 1686.

Oreskes, N. and Conway, E. (2010). *Merchants of Doubt: How a handful of scientists obscured the truth on issues from tobacco smoke to global warming.* London: Bloomsbury.

Orr, D. (1994). *Earth in Mind*. Washington, DC: Island Press.

Overland, I. and Sovacool, B. (2020). The misallocation of climate research funding. *Energy Research & Social Science*, 62, 101349.

Ozawa-Meida, L., Brockway, P., Letten, K., Davies, J. and Fleming, P. (2013). Measuring carbon performance in a UK University through a consumption-based carbon footprint: De Montfort University case study. *Journal of Cleaner Production*, 56, 185–98.

Pateman, C. (1970). *Participation and Democratic Theory*. Cambridge: Cambridge University Press.

Pedrosa, R. and Knobel, M. (2023). Higher education in Brazil and the new administration: What is at stake? *International Higher Education*, 116, 19–21.

Perales Franco, C. and McCowan, T. (2021). Rewiring higher education for the Sustainable Development Goals: The case of the Intercultural University of Veracruz, Mexico. *Higher Education*, 81, 69–88.

Pérez Salgado, F., de Kraker, J., Boon, J. and Klink, M. van der (2012). Competences for climate change education in a virtual mobility setting. *International Journal of Innovation and Sustainable Development*, 6(1), 53–65.

Perkin, H. (2007). History of universities. In J. J. F. Forest and P. G. Altbach (eds), *International Handbook of Higher Education,* 159–206. Dordrecht: Springer.

Peterson, J. B. (2018). *12 Rules for Life: An antidote to chaos*. London: Penguin.

Phelan, L. and Lumb, M. (2021). Higher education for times of climate crisis – critical awareness, purpose and community. *International Studies in Sociology of Education*, 30(1–2), 173–90.

Pinker, S. (2018). *Enlightenment Now: The case for reason, science, humanism, and progress*. London: Penguin.

Plymouth University (2015). 7 Steps to Taking an Institutional Approach to Learning about Sustainability. Available at: https://www.plymouth.ac.uk/rails/active_storage/blobs/proxy/eyJfcmFpbHMiOnsibWVzc2FnZSI6IkJBaHBBazlWIiwiZXhwIjpudWxsLCJwdXIiOiJibG9iX2lkIn19--8ae53af1c52a8830b6a53e4e7b8dc57a917941f6/596109_261397__Steps_to_Taking_an_institutional_approach_to_learning_about_sustainability_bold.pdf. Accessed 30 May 2024.

Polanyi, M. (2009). *The Tacit Dimension*. Chicago: University of Chicago Press.

Polsky, C. (2019). Open records, shuttered labs: ending political harassment of public university researchers. *UCLA Law Review*, 14 December, 208–93.

Pope Francis (2015). *Laudato Si': On care for our common home*. Huntington: Our Sunday Visitor.

Powell, J. (2019). Scientists reach 100% consensus on anthropogenic global warming. *Bulletin of Science, Technology & Society*, 37(4), 183–4.

Preston, J., Chadderton, C., Kitagawa, K. and Edmonds, C. (2015). Community response in disasters: An ecological learning framework, *International Journal of Lifelong Education*, 34(6), 727–53.

Purcell, W., Henriksen, H. and Spengler, J. (2019). Universities as the engine of transformational sustainability toward delivering the sustainable development goals. 'Living labs' for sustainability, *International Journal of Sustainability in Higher Education*, 20(8), 1343–57.

Qiang, Z. (2003). Internationalization of higher education: Towards a conceptual framework. *Policy Futures in Education*, 1(2), 248–70.

Rapley, C. G., de Meyer, K. and Chaytor, S. (2014). *Time for change? Climate science reconsidered.* Report of the UCL Policy Commission on Communicating Climate Science. London: UCL.

Rapley, C. (2023) Paper presented at the 'Higher Education and the Climate Crisis' conference, Senate House, London, 19 October.

Read, B. (2023). The university as heterotopia? Space, time and precarity in the academy. *Access: Critical Explorations of Equity in Higher Education,* 11(1).

Readings, B. (1996). *The University in Ruins*. Cambridge, MA: Harvard University Press.

Redfern, P. and Zhong, H. (2017). The role of EcoCampus in addressing sustainability in UK universities. *Frontiers of Engineering Management*, 4(2), 193.

Reid, A. (2019). Climate change education and research: Possibilities and potentials versus problems and perils? *Environmental Education Research*, 25(6), 767–90.

Reimers, F. M. (2021). *Education and Climate Change: The role of universities*. New York: Springer Nature.

Rensimer, L. and McCowan, T. (2023). *Reconceptualising International Flows of Aid to and through Higher Education*. Centre for Global Higher Education, Working Paper No. 97.

Rittel, H. and Webber, M. (1973). Dilemmas in a general theory of planning. *Policy Sciences*, 4, 155–69.

Robinson, K. S. (2021). *The Ministry for the Future*. New York: Little, Brown and Co.

Robinson, O. J., Tewkesbury, A., Kemp, S. and Williams, I. D. (2018). Towards a universal carbon footprint standard: A case study of carbon management at universities. *Journal of Cleaner Production*, 172, 4435–55.

Robinson-Pant, A. (2020). Inclusive education: Thinking beyond systems. *Compare: A Journal of Comparative and International Education*, 50(5), 619–38.

Rolleston, C. et al. (2023). *Stepping Up or Falling Behind? Students' views on universities and the climate crisis*. Transforming Universities for a Changing Climate, Working Paper Series No. 10.

Rooney, M. and McMillin, J. (2010). The campus as a classroom: Integrating people, place and performance for communicating climate change. In W. Leal Filho (ed.), *Universities and Climate Change: Introducing climate change to university programmes*, 117–36. Heidelberg: Springer.

Rousell, D. and Cutter-Mackenzie-Knowles, A. (2020). A systematic review of climate change education: Giving children and young people a 'voice' and a 'hand' in redressing climate change. *Children's Geographies*, 18(2), 191–208.

Royal Anniversary Trust (2023). *Accelerating the UK Tertiary Education Sector towards Net Zero: A sector-led proposal for action and connected thinking*. Available from: https://www.universitiesuk.ac.uk/topics/climate-and-sustainability/accelerating-uk-tertiary-education. Accessed 1 February 2024.

Rumbley, L. E. (2020). Internationalization of higher education and the future of the planet. *International Higher Education*, 100, 32–4.

Ryle, G. (1949). *The Concept of Mind*. London: Penguin Books.

Salinas, I., Guerrero, G., Satlov, M. and Hidalgo, P. (2022). Climate change in Chile's school science curriculum. *Sustainability*, 14(22), 15212.

Salto, D. and Levy, D. (2021). For-profit higher education in Latin America: Exception or precursor? *International Higher Education*, 106, 32–4.

Santos, B. de S. (2004). *A universidade do século XXI: Para uma reforma democratica e emancipatória da universidade*. São Paulo: Cortez.

Santos, B. de S. (2015). *Epistemologies of the South: Justice against epistemicide*. New York: Routledge.

Santos, B. de S. (2017). *Decolonising the University: The challenge of deep cognitive justice.* Newcastle: Cambridge Scholars Publishing.

Sasidevan, D. and Santha, S. D. (2023). *Fostering Adaptive Innovation and Climate Action: A participatory research study involving students, social entrepreneurs and community actors*. Transforming Universities for a Changing Climate, Working Paper Series No. 17.

Sauvé, L. (1996). Environmental education and sustainable development: A further appraisal. *Canadian Journal of Environmental Education*, 1, 7–34.

Schendel, R. (2015). Critical thinking at Rwanda's public universities: Emerging evidence of a crucial development priority. *International Journal of Educational Development*, 42, 96–105.

Schendel, R., McCowan, T., Rolleston, C., Adu-Yeboah, C., Omingo, M. and Tabulawa, R. (2023). Pedagogies for critical thinking at universities in Kenya, Ghana and Botswana: The importance of a collective 'teaching culture'. *Teaching in Higher Education*, 28(4), 717–38.

Schendel, R., DeLaquil, T., Rensimer, L. and McCowan, T. (2024) 'Aid' to higher education in a multiplex world: Exploring the diversity of donor discourses and rationales. *International Journal of Educational Development*, 111(2), 103162.

SEMESP (2024). Pesquisas apontam crescimento de alunos no presencial e queda no valor das mensalidades. SEMESP, 25 March. Available at: https://www.semesp.org.br/noticias/pesquisa-aponta-crescimento-de-alunos-no-presencial-e-queda-no-valor-das-mensalidades/. Accessed 24 May 2024.

Senbel, M., Ngo, V. D. and Blair, E. (2014). Social mobilization of climate change: University students conserving energy through multiple pathways for peer engagement. *Journal of Environmental Psychology*, 38, 84–93.

Sessions, G. (1987). The deep ecology movement: A review. *Environmental Review*, 11(2), 105–25.

Shahjahan, R. A. (2015). Being 'lazy' and slowing down: Toward decolonizing time, our body, and pedagogy. *Educational Philosophy and Theory*, 47(5), 488–501.

Shattock, M. (2009). *Entrepreneurialism in Universities and the Knowledge Economy: Diversification and organisational change in European higher education*. Bletchley, UK: Open University Press.

Shields, R. (2019). The sustainability of international higher education: Student mobility and global climate change. *Journal of Cleaner Production*, 217, 594–602.

Shields, R. and Lu, T. (2023). Uncertain futures: Climate change and international student mobility in Europe. *Higher Education*, 19 April, 1–18.

Shiva, V. (1993). *Monocultures of the Mind*. London: Zed Books.

Silova, I. (2021). Facing the Anthropocene: Comparative education as sympoiesis. *Comparative Education Review*, 65(4), 587–616.

Singh, M. (2012). Re-inserting the 'public good' into higher education transformation. In B. Leibowitz (ed.), *Higher Education for the Public Good–Views from the South*, 1–16. Stoke on Trent, UK: Trentham Books; and Stellenbosch: Sun Media.

Stein, S. (2019). The ethical and ecological limits of sustainability: A decolonial approach to climate change in higher education. *Australian Journal of Environmental Education*, 35(3), 198–212.

Stein, S. et al. (2023). Beyond colonial futurities in climate education. *Teaching in Higher Education*, 28(5), 987–1004.

Stephens, J. C., Frumhoff, P. C. and Yona, L. (2018). The role of college and university faculty in the fossil fuel divestment movement. *Elementa: Science of the Anthropocene*, 6, 41.

Sterling, S. (2011). Transformative learning and sustainability: Sketching the conceptual ground. *Learning and Teaching in Higher Education*, 5(11), 17–33.

Sterling, S. and Maxey, L. (2013). Introduction. In Sterling, S., Maxey, L. and Luna, H. (eds), *The Sustainable University. Progress and Prospects.* Abingdon: Routledge.

Stevenson, R. B., Nicholls, J. and Whitehouse, H. (2017). What is climate change education? *Curriculum Perspectives*, 37(1), 67–71.

Stokes, D. E. (1997). *Pasteur's Quadrant: Basic science and technical innovation*. Washington, DC: Brookings Institution Press.

Students Organising for Sustainability UK (SOS-UK) (2021). *Student Opinion – Climate Change and COP26*. Available at: https://assets-global.website-files.com/6008334066c47be740656954/61f7aed2c46c9a48ce55ee93_20220131_SOS-UK_Student%20Opinion%20COP26.pdf. Accessed 23 May 2024.

Tannock, S. (2021). *Educating for Radical Social Transformation in the Climate Crisis*. London: Palgrave Macmillan.

Thunberg, G. (ed.) (2024). *The Climate Book*. London: Penguin.

Tikly, L. (2019). *Education for Sustainable Development in the Postcolonial World: Towards a transformative agenda*. London: Routledge.

Tomlinson, M. (2012). Graduate employability: A review of conceptual and empirical themes. *Higher Education Policy*, 25, 407–31.

Torres-Olave, B. and Bravo González, P. (2021). Facing neoliberalism through dialogic spaces as sites of hope in science education: experiences of two self-organised communities. *Cultural Studies of Science Education*, 16, 1047–67.

Treisman, R. (2021). Harvard University will stop investing in fossil fuels after years of public pressure. NPR, 10 September. Available at: https://www.npr.org/2021/09/10/1035901596/harvard-university-end-investment-fossil-fuel-industry-climate-change-activism. Accessed 30 May 2024.

Trencher, G., Yarime, M., McCormick, K., Doll, C. and Kraines, S. (2014). Beyond the third mission: Exploring the emerging university function of co-creation for sustainability. *Science and Public Policy*, 41(2), 151–79.

Tsing, A. L. (2015). *The Mushroom at the End of the World*. Princeton: Princeton University Press.

UNDP and University of Oxford (2021). Peoples' Climate Vote: Results. Available at: https://www.undp.org/sites/g/files/zskgke326/files/publications/UNDP-Oxford-Peoples-Climate-Vote-Results.pdf. Accessed 4 November 2024.

UNESCO (2006). *Framework for the UNDESD International Implementation Scheme*. Paris: UNESCO. https://unesdoc.unesco.org/ark:/48223/pf0000148650. Accessed 4 November 2024.

UNESCO (2019). *Country Progress on Climate Change Education, Training and Public Awareness: An analysis of country submissions under the United Nations Framework Convention on Climate Change*. Paris: UNESCO.

UNESCO IESALC (2022). *Exploring International Aid for Tertiary Education: Recent developments and current trends*. Available at: https://unesdoc.unesco.org/ark:/48223/pf0000381747. Accessed 1 June 2023.

UNESCO Institute for Statistics (UIS) (2022). Gross enrolment ratio for tertiary education. Available at: http://data.uis.unesco.org/. Accessed 1 February 2022.

UNESCO Institute for Statistics (UIS) (2023a). Gross enrolment ratio for tertiary education. Available at: http://data.uis.unesco.org/. Accessed 1 September 2023.

UNESCO Institute for Statistics (UIS) (2023b). Government expenditure per student: tertiary education. Available at: http://data.uis.unesco.org/. Accessed 1 February 2024.

United Nations (1992). *United Nations Framework Convention on Climate Change* (FCCC/INFORMAL/84 GE.05-62220 (E) 200705). Available at: unfccc.int/resource/docs/convkp/conveng.pdf. Accessed 13 December 2024.

Unterhalter, E., Allais, S., Howell, C., McCowan, T., Morley, L., Oanda, I. and Oketch, M. (2017). *Higher Education and the Public Good: Concepts, challenges and complexities in Africa*. Working paper for the research project, Higher Education, Inequality and the Public Good in Four African Countries.

Vaughter, P., Wright, T., McKenzie, M. and Lidstone, L. (2013). Greening the ivory tower: A review of educational research on sustainability in post-secondary education. *Sustainability*, 5(5), 2252–71.

Wesselink, A. and Gouldson, A. (2014). Pathways to impact in local government: The mini-Stern review as evidence in policy making in the Leeds City Region. *Policy Sciences*, 47, 403–24.

White, J. (2019). The end of powerful knowledge? *London Review of Education*, 17(3), 429–38.

Williams, B. (2001). The theoretical links between problem-based learning and self-directed learning for continuing professional nursing education. *Teaching in Higher Education*, 6(1), 85–98.

Williamson, B. (2018). The hidden architecture of higher education: Building a big data infrastructure for the 'smarter university'. *International Journal of Educational Technology in Higher Education*, 15(1), 1–26.

Willis, P. (1977). *Learning to Labour: How working class kids get working class jobs*. Bournemouth: Saxon House.

Winter, J. and Cotton, D. (2012). Making the hidden curriculum visible: sustainability literacy in higher education. *Environmental Education Research*, 18(6), 783–96.

World Bank (2024). Education Statistics – All Indicators. Available at: https://databank.worldbank.org/indicator/SE.TER.ENRL?id=c755d342andreport_name=EdStats_Indicators_Report&populartype=series. Accessed 19 May 2024.

Wynveen, C. J., Kyle, G. T. and Tarrant, M. A. (2012). Study abroad experiences and global citizenship: Fostering pro-environmental behavior. *Journal of Studies in International Education*, 16(4), 334–52.

Young, M. F. D. (2008). *Bringing Knowledge Back In: From social constructivism to social realism in the sociology of education*. London: Routledge.

# Index

Figures have *italic* page numbers while tables are in **bold**. Notes are indicated by 'n' after the page reference.


academic disciplines, awareness of 117–18
actions (engagement framework) 189
activists, climate 11, 17, 37, 40, 46, 107, 186, 203
 *See also individual groups*
actors (engagement framework) 188–9
actors, bridging 64–5, *64*, *67*, 77, 98, 192, 193, 194
adaptation 20–1, 26, 41, 44, 97, 119, 171
adaptation pathways, mitigation and 66–73, *67*, **68–9**, 79n3
'adjectival' education 108
aerosols, use of 23
Africa 4, 97, 179
agency (ownership) 141
agriculture 18, 143, 179, 180, 191–2, 203
Alexander, Bryan, *Universities on Fire: Higher education in the climate crisis* 6, 172
Almassi, B. 37
'alternative' truths 34
Alves, F. 20
Amazon rainforest 194
Anthropocene period 18, 30n1, 108
antiharassment policies 158
application (connection) 142
Arizona State University 154
Arrhenius, Svante 18
arts, the 78, 112, 115, 120
assemblages (definition) 197n3
Association of University Leaders for a Sustainable Future 2, 101
Atwood, Margaret, *Oryx and Crake* 199
Australia 83, 136, 158, 194
*autopoiesis* (definition) 23, 74
availability (access) 141
*Avengers: Infinity War* (Marvel 2018 film) 199
awareness, climate change 31–53
axiological concerns (human enquiry) 113, 118–21, 127

Bacevic, J. 197n3
Bacon, Francis 114
Bangladesh 83
Barcelona, University of 186
Barth, M. 49–50, 138
Bell, Alice 39
Berners-Lee, M. 23
Bernstein, Basil 61
Binagwaho, A. 128, 193, 194
biodiversity, loss of 18, 19
biometeorology 138
Booker, Stephen 39, 40
Booth, A. 183
Boslough, M. 40
Branson, Richard 25–6, 42
Brazil 84, 88, 158, 183–4
Brexit campaign 34
bridging actors 64–5, *64*, *67*, 77, 98, 192, 193, 194
Bristol, University of 160–1
Brugger, J. 181, 191–2, 193
bullying policies 158
Burandt, S. 49–50, 138

Cambridge, University of 171
campus operations
 engaging with society 191
 five modalities of the university 56, 58–9, *59*, 60, 61
 internationalisation of higher education 95, 101n2
Canada 83
capitalism 21, 28, 33, 34
Capra, Fritjof 114
Capstick, S. 161
carbon absorption 112
carbon capture 24, 26, 41, 120, 169
carbon dioxide 18, 41, 89, 97
 reducing emissions 24, 112, 159, 186
 student flight emissions 89, 91, 155–6, 158
carbon footprint
 campuses 2
 livestock 158
 procurement policies 157–8
carbon neutrality 2, 13n3
Carson, Rachel, *Silent Spring* 19
cars, petrol/diesel 27
Casey, O. 107, 108, 127
catalysing embodiment 170–3
catastrophes, human 15
Centre for Climate Change and Sustainability Education (UCL) 184
certification (ownership) 142
China 84, 100

chlorofluorocarbons 18
citizenship education 50–1
civilisational crisis, climate change as a 15–30
 definition 16
'cli-fi' genre literature 115
Climate Action Network for International Educators (CANIE) 91
climate change
 *anthropogenic* nature 17, 21–2
 basics 17–21
 broader concerns and 19
 as a civilisational crisis 15–30
 definition 17–18
 emotional dimension 35
 internationalisation of higher education and 6–7, 89–92, 94, 95, 96–7
 its *complexity* 17, 23–4
 its *contested* nature 17, 24–5, 34
 its *global* reach 17, 25–6
 its *time-bound* nature 17, 27–8
 key characteristics of 21–8, 29
 materiality of 3–4
 Stages of impact of the university on *64*
'Climategate' incident (2009) 37
#climatescam 37–8
climate science
 denial, distraction and delay 37–41
 is there a consensus? 36–7
 understanding climate 36–41
climate system 114
Climate-U project 5, 10, 13n4, 62, 135, 170, 194
Coady, D. 38, 39
coal 18
collaboration (connection) 142
collaboration (engagement framework) 189–90
collective knowledge 1, 32, 59
colonialism 34
'community' (definition) 177
community engagement 58, 61, 175–97
community of social networks (CSN) 171
competences, defining 49–50
*Concept of Mind, The* (Ryle) 153
Conferences of Parties (COPs) 22
'connected curriculum' framework 62, 143, 168
*Connected Curriculum* (Fung) 62–3, 138, 168
consensus, climate science 36–7
Conservative government, UK 27
consumer choices 21, 111, 125, 161
continuity (access) 141
Conway, Eric, *Merchants of Doubt* 36, 39, 73
Cook, J. 36
Cooperative Extension System, US 191
co-production 44–5, 62, 93, 169, 178, 181, 190, 192–5
Cordero, E. C. 97, 107
Córdoba, National University of (Argentina) 179
Córdoba reforms 4
Corry, R. 38, 39
Cotton, D. 162
courses, climate change 132
Covid-19 pandemic 26, 34
Cox, Brian 65
creative unruliness (transformation) 204–5
Crimmins, M. 181, 191–2, 193
critical questioning (pedagogical foundations) 122–5, 127
Croatia 89
Crow, Michael 154
cultural studies 115
curricular features 140–3
 12 curricular 'features' 141–2
 access 141
 connection 142
 ownership 141–2
 topographical features in four curricular interventions 143, **144–5**
curriculum
 climate change in the formal 135–8
 'connected curriculum' framework 62, 143, 168
 hidden 132, 134, 161–4, 166
 'teacher-proof' 151
 topography in the university *139*

Dar es Salaam, University of 175
Dawkins, Richard 65
deforestation 19
deliberation (pedagogical foundations) 125–7
denial, distraction and delay, climate 37–41
 scepticism and 11, 38, 39
 'six stages' 38
 types of 39, 41
deniers, climate change 116, 126, 167
 *delay* not *denial* 27
 'full-fat' denial 39
 global warming 23, 24–5
Denmark 127, 194
desertification 19
'design elements' (climate change education) 107
'developmental' model, university 87
'developmental university' (Africa) 179
Development Studies, University for (Ghana) 183
digital technologies, use of 107, 110, 116, 206
disciplinarity (connection) 142
disciplinary divide 43–4
disembodied knowledge 182
disparities, resource 83–5
distance programmes 98
diverse knowledge traditions 44–5
diversity (university–society engagement) 192
 divestment 160, 185–6
dynamics of higher education, global 81–101
 global influences 82–8

Earth Summit (Rio de Janeiro, 1992) 18–19
East Anglia, University of 37
East Asia 88
ecology
 interdependence in 114
 of knowledges 32, 45, 93, 118, 128, 193
economics (academic discipline) 117, 138
eco-pedagogy movement 125
'ecoversities' 172–3
Ecoversities Alliance 205
*Educating for Radical Social Transformation in the Climate Crisis* (Tannock) 7
educational responsibilities, broader 111
electricity 18, 156, 157

embeddedness (connection) 142
embodied knowledge 182
embodiment, Institutional 153–73
 catalysing embodiment 170–3
 in research agendas 166–70
 prefiguring the climate transformation 155–65
*emergence* (definition) 23
emissions
 institutional 95, 156–7, 171
 reducing 24, 112, 157, 159, 186
 student flight 89, 91, 155–6, 158
 *See also* greenhouse gases
employability 86, 87, 88, 111, 178–9
energy consumption, increased 23
energy use, efficient 17
'engagement' (definition) 177
engaging with society 175–97
 evolution of the third pillar 178–82
 framework of university engagement with society 187–92
 limits of community engagement 195–6
 services 182–4
 six main dynamics 187–90, *190*
Enlightenment, the 34, 110, 112
'enterprise' (community engagement) 181
environmental education, older tradition of 108
environments, learning 161–4
epistemic backdrop 32–5
'epistemic humility' 44, 207
epistemic pluralism 45, 110, 118, 128, 150, 169, 188, 194, 195
epistemological concerns (human enquiry) 113–18, 127
epistemology 169
equity, diversity and inclusion (EDI) 158, 161
ethics, climate 87, 120, 168–9
European Union (EU) 12n2, 40, 84
European universities 133
Evans, J. 165
evolution of the third pillar 178–82
experientiality (connection) 142
experimentation (institutional embodiment) 164–5
'extension' (community engagement) 180, 184
extinction 31
Extinction Rebellion 150

Facer, Keri 5–6, 20, 52–3, 143, 148
Fahey, S. J. 150
Fals Borda, Orlando 194
fertilisers, chemical 19
Findler, F. 56, 76
five modalities of the university 56–63, *59*
 campus operations 56, 58–9, *59*, 60, 61
 education 56, 57, 58, *59*, 60, 65
 four kinds of relationship 62
 impact of 63
 knowledge production 56, 57, *59*, 60
 mitigation and adaptation pathways *67*
 public debate 56, 57–8, *59*, 61
 services 56, 57, 58, *59*, 60
'flipped classroom' 111, 126, 129n1
flow (engagement framework) 188
Folkehøjskoler (folk high schools), Denmark 127, 194
Foote, Eunice 18, 169
forge of ideas (transformation) 202–3
for-profit institutions 3, 83, 88, 153, 171
fossil fuels
 alternatives to 17
 burning 18, 41
 companies 160
 *delay* not *denial* 27
 divestment 160, 185–6
 exploitation by lobby 37
 industry development 119
 reducing company use 183
fossil gas 18, 30n2
framework of university impact on climate change 63–78
France 84
Francis, Pope, *Laudato Si'* 120, 168
Frediani, A. 6
Freedom
 academic 134, 151, 169, 185, 186, 200, 204
 individual 27, 35, 41, 119–20
Freire, Paulo, *Pedagogy of the Oppressed* 9, 108, 123–5, 129, 163, 169–70, 194
Fricker, M. 45
Fridays for Future 150, 185
funding, research 2, 44, 166–7
Fung, Dilly, *Connected Curriculum* 62–3, 138, 168

gas, fossil 18, 30n2
gas industry 175–6
Gates, Bill, *How to Avoid a Climate Disaster* 42, 180
geo-engineering 120, 132, 159
 interventions 23, 24, 25
Germany 49, 84
Ghana 183
Ghosh, Amitav, *The Great Derangement* 31, 33, 35, 115
Global Alliance of Universities on Climate 101
global dynamics, navigating 97–101
 of higher education 81–101
global influences
 and the ways universities operate 82–8
Global North 7, 26, 27, 42, 84, 93, 101, 119, 128
Global South 7, 84, 93, 119, 128, 209
good life, the 120
'the good life' 34–5
Gore, Al, *Inconvenient Truth* 24
Gouldson, A. 183
Gove, Michael 34
government investment
 for universities 3
Grady-Benson, J. 185–6
*Great Derangement, The* (Ghosh) 31
greenhouse gases 3–4, 15–16
 emissions 47, 73, 91, 155–6
 fossil fuels and 18
 reducing 48, 157
 solutions to 17
Green New Deal policy framework 159

Hansen, James 18
Harvard University 171, 186
Haydon, G. 151

health clinics 183
Helm, D. 112
Henderson, J. 56
'hermeneutical' forms of knowledge 45
hidden curriculum 132, 134, 161–4, 166
*Higher Education for and beyond the Sustainable Development Goals* (McCowan) 79, 163, 169
Higher Education (Freedom of Speech) Act (2023) 197n2
Higher Education institutions (HEIs) 3, 26, 42–3, 55, 86–7, 133, 151–2, 156, 205
Hillmer-Pegram, K. C. 183
history, teaching of 120
'hockey stick' study 40, 75
hospitals 55, 60, 94, 183
*How to Avoid a Climate Disaster* (Gates) 33, 35, 115, 180
human beings, ontological reflection and 114–15
human, enquiry, three dimensions of 113–21
  axiological 113, 118–21, 127
  epistemological 113–18, 127
  ontological 113, 114–15, 127
humanities 44, 78, 112, 115, 120
human rights education 50–1
'hyperporosity' (definition) 188, 196
hypocrisy (institutional embodiment) 160, 196

impact, direct and indirect 76–7
impact, modalities and pathways of 55–79
  'bridging actors': 64, *64*, 65, *67*, 77
  ecosphere *64*, 65–6, *67*, 77
  five modalities of the university 56–63, *59*
  a framework of university impact on climate change 63–77, *64*
  negative impacts of universities on climate change **72**
  positive impacts of universities on climate change **70–1**
  society 64–6, *64*, *67*, 77
impact on climate change, a framework of university 63–77
  characteristics of the model 73–7
  mitigation and adaptation pathways 66–73
impacts, dynamics of
  intensity 97–8
  reach 97, 98
  through internationalisation 99, *99*
  timescale 97, 98
impact study (McCowan) 75
implications for university practice 28–30, 148–52
  criticality 149
  epistemic pluralism 150
  non-coercion 149–50
  students' existing knowledge 150
income, country and
  high 21, 85, 109
  low 83, 84, 119, 209
  lower 21, 85
  lower-middle 84, 109
  low-income 109
  upper-middle 109
income-generating activities 3
*Inconvenient Truth* (Gore) 24
India 25, 62, 84, 127
Indigenous knowledge 45, 118, 164, 169, 194
Indigenous peoples 44, 194, 197n1
individual knowledge 59
industrialisation 7, 18, 21, 27, 33, 201
Industrial Revolution 4, 18, 27, 119
influence
  activity examples **72**
  adaptation pathways **69**
  areas of activity **70–1**
  Mitigation and adaptation pathways for university influence on climate change *67*
  mitigation pathways **68**
  outside the institution 4
information brokers 192
injustices, climate 25, 113, 119, 123
institutional embodiment 153–73
institutional emissions 95, 171
  Scope 1, 2 and 3 95, 156–7, 171
interdisciplinarity work 137, 138
Intergovernmental Panel on Climate Change (IPCC) 17, 27, 36–7, 168, 194
intermodality 63
international branch campuses (IBCs) 91, 95, 98
internationalisation of higher education 88–97
  actors 90–1, 93
  campus operations 95, 101n2
  dynamics of impact through *99*
  education modality 91–2, 101n2
  influence 90–1, 93
  Internationalising the education modality 96–7, *96*
  knowledge production 92–3
  practice 90–1, 92, 93
  public debate 94
  services 94
*International Journal of Sustainability in Higher Education* 6
international treaties 33
International Universities Climate Alliance 2
international university rankings 43, 86
'inverted classroom' 126
investment portfolios 160
Italy 105

Jamaica 89
JBSE (justice-based sustainability education) 128
Jickling, B. 121
Jimenez, J. 128
John Hopkins University 26
Just Stop Oil 25, 186

Kabachnik, P. 128
Kerr, Clark 55, 94
Kinol, A. 159
Klein, Naomi 112–13
knowledge 41–53
  acquiring and using knowledge 46–53
  and climate change 31–53
  collective 1, 32, 59
  definition 47
  different languages and 49
  ecology of knowledges 32, 45, 93, 118, 128, 193

embodied/disembodied 182
generating knowledge 42–5
generation of 3, 133
Indigenous 45, 118, 164, 169, 194
individual 59
networks 192
production 56, 57, *59*, 60, 92–3
society 180
students' existing 150
'transfer' 181
Koch family 167
König, A 165
Koonin, Steven, *Unsettled* 39–40
Kwauk, C. 107, 108, 127

languages
English as lingua franca 94, 101n4
English-speaking countries 93
global/local audiences 100
knowledge and different 49
major higher education 85
Latin America 4, 88, 179, 180, 194
Latter, B. 161
*Laudato Si'* (Pope Francis) 73, 120
Leal Filho, W. 6, 168
learning experience, providing a 161–4
Leeds City Council 183
Lee, J. J. 95
legal clinics 55, 57, 60, 98, 189,
Lehtonen, A. 114
Levin, K. 19
literature, on climate action 5–7, 104, 115
'living lab' (definition) 164–5
local knowledge 45
Lomborg, Bjorn 39
London School of Economics and Political Science (LSE) 2
Lotz-Sisitka, H. 6
Lovelock, James 114
Lubchenco, J. 167–8
Lundemo, O. A. 95
Luxembourg 83

McCowan, T. 56, 59, 75
*Higher Education for and beyond the Sustainable Development Goals* 79, 163, 169
Macintyre, T. 6
malleability (ownership) 141
management, university 170–2
key tasks of leaders 170–1
Mann, Michael E. 38, 39, 40, 75, 187
MAPA (Most Affected People and Areas) 119 11
Marginson, S. 57–8
Margulis, Lynn 114
marketisation 2–3, 168, 179
Marshall, George 40–1, 46–7, 187
massification 109, 110, 134
massive open online courses (MOOCs) 79n1, 110, 147
Mastercard Foundation 84
Maxey, L. 172
media 57, 65, 185, 186, 187, 188
media studies 115
medicine (academic discipline) 166
*Merchants of Doubt* (Oreskes and Conway) 36, 39, 73
methane 18
methodology 40, 86, 108, 169, 178, 194
Mexico 105, 194, 205
Millennium Development Goals (MDGs) 137
*Ministry for the Future, The* (Robinson) 25
Misiaszek, G. W. 128
mitigation 20–1, 44, 97, 119, 168, 171
and adaptation pathways 66–73, *67*, **68–9**, 79n3
mobility, student/staff 29, 83, 88, 89–90, 91, 93, 95, 101, 101n2, 157
modalities and pathways of impact 55–79
'bridging actors': 64, *64*, 65, *67*, 77
ecosphere *64*, 65–6, *67*, 77
five modalities of the university 56–63, *59*
a framework of university impact on climate change 63–77, *64*
negative impacts of universities on climate change **72**
positive impacts of universities on climate change **70–1**
society 64–5, *64*, 65–6, *67*, 77
modalities, university *59*, *96*
Molthan-Hill, P. 6, 136
'monoculture of the mind' 201
Monroe, M. 106, 127
Morano, Marc 39, 40
Morrill Act, USA (1862) 179
Mozambique 100
multidisciplinarity work 137, 138
'multiversity' 55, 94

natural gas 30n2
natural sciences 112
natural world 4, 22
Nepal 85
net zero carbon 2, 13n3
nitrous oxide 18
non-Western knowledge 44
'no-platforming' (definition) 186
'no regrets' policies 112
Norgaard, K. M. 41, 47
North America 181
North American Arctic 194
Northern British Columbia, University of 183
Norwegian Programme for Capacity Development in Higher Education and Research for Development (NORHED) 85
Nussey, C. 6

oceanography 138
oil 18, 175–6
Ojala, M. 48–9
online learning 29, 91, 95, 147, 191
*See also* massive open online courses (MOOCs)
ontological concerns (human enquiry) 113, 114–15, 127
Oreskes, Naomi, *Merchants of Doubt* 36, 39, 73
Organisation for Economic Cooperation and Development (OECD)
Creditor Reporting System 84
PISA 132, 152n1
orientation (engagement framework) 189

*Oryx and Crake* (Atwood) 199
'Our' (concept) 16
'outreach' (community engagement) 180–1
Overland, I. 44
Oxford, University of 26, 33, 38, 171, 186

palaeoclimatology 132, 138
Paris Agreement (2015) 1, 35
participatory action research 62, 93, 100, 181, 188, 194
pedagogical foundations 121–7
  critical questioning 122–5, 127
  deliberation 125–7
pedagogical renewal, towards 128–9
*Pedagogy of the Oppressed* (Freire) 168
Penn State University 183
pesticides 19
'phenomena-based' learning 114
PISA (Programme for International Student Assessment) 152n1 132
pluralism, epistemic 45, 110, 118, 128, 150, 169, 188, 194, 195
'pluriversities' 205
Plymouth, University of 2, 162
politics
  political conflicts 27
  political elite 34
  political questioning 123–4
  presidential election (2016) 34
  values 48–9, 112
pollution levels 19, 189–90
populist movements (2016) 110, 112
porosity (engagement framework) 187–8, 192, 195
Powell, J. 36
prefiguring the climate transformation 155–65
  acting as a space for experimentation 164–5
  climate action in the here and now 155–9
  providing a rich learning experience 161–4
  serving as an exemplar 159–61
private good 43, 60, 79
problem-based learning 123–4
procurement policies 157–8, 161
protagonist, university as (transformation) 206–7
public communication 40, 41, 46–7, 65
public debate 186–7
'public engagement' 181
public good 3, 28, 43, 60, 79, 84, 101, 182, 196, 208
public information 26
public policy 138

quadruple helix model 180

rankings, international university 43
Rapley, C. 167–8, 177, 187
Readings, Bill, *The University in Ruins* 206
recruitment, student 2
regeneration 26, 41, 44, 97, 119, 171
  definition 20–1
reimagining institutions (transformation) 205–6
Reimers, F. M. 6, 146
religions, major 120, 158–9
renewable energy 27
  alternatives 97–8
research agendas, embodiment in 166–70
research quality 110
resource disparities, global 83–5
resourcing, increase in 83
response of education to the climate crisis 105–8
retrofitting buildings 157
Rhodes Must Fall movement 110, 118
Robinson, Kim Stanley, *The Ministry for the Future*, 25
Robinson-Pant, A. 78
Rodrigues, C. 128
Royal Anniversary Trust (2023) report 156–7
Russell Group universities 91
Ryle, Gilbert, *The Concept of Mind* 153

Santos, B. de S. 61, 205, 206
São Paulo, University of 138
Sarathy, B. 185–6
scepticism 11, 38–9, 116, 128, 134, 163, 184–5
schools
  assemblies 163
  climate change education 6, 51, 104, 151
  curriculum topography 131, 133–4, 146, 161
  engagement with 184
  Fridays for Future 150, 185
  institutional embodiment 155
  quality of schooling 109
  staff 110
science, technology, engineering and mathematics (STEM) 44, 53n1, 112, 120, 166, 168
scientists, climate 17, 21, 28, 35–6, 39–40, 46–7, 116–17, 168, 187
Scope 1, 2 and 3 (institutional GHG classification) 95, 156–7, 171
'semi-skimmed' denial 39
senior management hierarchy (SMH) 171
services (engaging with society) 182–4
'7 Steps to Taking an Institutional Approach to Learning about Sustainability' (Plymouth University) 162
Shields, R. 89, 155
*Silent Spring* (Carson) 19
sites of learning 140, 152n3
skills
  essential for addressing sustainable development 48
  hard and soft 47–8, 49
  Is it the role of the university to teach about climate change? 133–4
'skimmed' denial 39
small is beautiful (transformation) 203–4
social enterprise 181
'socially organized denial' 41
social media engagement 100, 107, 116, 147
social media, rise of 34
social sciences 44, 78, 112, 120
society, engaging with 175–97
  evolution of the third pillar 178–82
  framework of university engagement with society 187–92

limits of community engagement 195–6
services 182–4
six main dynamics 187–90, *190*
Socrates 122–3
Songo Songo Island 175
South Africa 100, 110, 118
South America 115
Southern Africa 115
Sovacool, B. 44
Stein, S. 6, 128
Sterling, S. 172
Strategic Partnerships for Higher Education Innovation and Reform (SPHEIR) 85
students
agency 205
existing knowledge 150
fees 83
flight emissions 89, 91, 155–6, 158
learning about climate change 51–2
mobility 29, 83, 88, 89–90, 91, 93, 95, 101, 101n2, 157
'subversities' 172–3, 205
Sulitest 47
*sumak kawsay* philosophy 115, 118
sustainability 170–1
co-creation for 193
factors involved in 26, 126, 154, 202
internationalisation of higher education and 84, 89, 91, 95, 101
land management 159
sustainable development, education for 49, 50–1, 108, 121
Sustainable Development Goals (SDGs) 2, 6, 73, 137
(Goal 13) 52–3, 105
Sustainable Development Solutions Network 2

Tannock, Stuart, *Educating for Radical Social Transformation in the Climate Crisis* 7
Tanzania 175
Tata Institute of Social Sciences, Mumbai 62
'teacher-proof' curriculum 151
teaching climate change
in the 'classroom' 103–29
five 'design elements' 107
'new green learning agenda' 108
pedagogical foundations 121–7
response of education to the climate crisis 105–8
teaching quality 110–11
three dimensions of human enquiry 113–21
topography approach 104
towards pedagogical renewal 128–9
trends in higher education as a learning space 109–13
Teaching Excellence Framework, UK 87
teaching-only institutions 110–11, 147
'tech transfer' 181
temperature, change in average 3, 18, 114
'testimonial' forms of knowledge 45
think tanks 39, 46, 73
third pillar, evolution of the 178–82, 189
'third stream' activities 179, 181
Tikly, L. 73
tipping points 27, 99, 114
Toles, T. 39, 40
topography
features in four curricular interventions **144–5**
role of the university to teach about climate change 104, 133–5
topography, curriculum 131–52
climate change in the formal curriculum 135–8
a framework of curriculum topography in the university 138–48, *139*
implications for university practice 148–52
in the university *139*
schools' curriculum 131, 133–4
Towards a European Framework for Community Engagement in Higher Education (TEFCE) Toolbox 197n4
traditions, knowledge 44, 45
transdisciplinarity work 137, 138
transformation, of universities 29–30, 199–209
creative unruliness 204–5
forge of ideas 202–3
reimagining institutions 205–6
small is beautiful 203–4
university as protagonist 206–9
transmodality 63, 142, 183, 189, 204
travel, international 5, 58, 73, 78, 89, 91, 93, 94, 95, 158
tree-planting schemes 19
Trencher, G. 193
trends in higher education as a learning space 109–13
triple helix model 179–80, 187
Tsing, Anna 114, 203
Turkey 84
Tyndall, John 18, 169

*ubuntu* philosophy 115, 118
UCL (University College London) 62, 157, 184
UK
Higher Education (Freedom of Speech) Act (2023) 197n2
Knowledge Exchange Framework 181
'third stream' activities 179
university fees/income 83, 158
UNESCO
BRIDGES programme 44
Institute for Statistics 83
United Arab Emirates (UAE) 84
United Nations Development Programme (UNDP) 33, 38
United Nations Framework Convention on Climate Change (UNFCCC) 8, 32, 49, 105, 132
UNITIERRA (Universidad de la Tierra, Mexico) 205
Universal Postal Union (UN agency) 209n1
universities, growth of 82–3
*Universities on Fire: Higher education in the climate crisis* (Alexander) 6, 172
university as protagonist (transformation) 206–7
*University in Ruins, The* (Readings) 206
university–society engagement, dynamics of *190*

*Unsettled* (Koonin) 39–40
USA
broader general curriculum 133
Cooperative Extension System 191–2
Indigenous institutions 194
'land-grant' universities 4, 179
for-profit institutions 88, 158
'third stream' activities 179
university fees and costs 83, 84
'usable science' 193

vaccine development 26, 64
values
embodiment of 148, 169
Is it the role of the university to teach about climate change? 134–5
political and moral 48–9, 112
Virgin Earth Challenge prize 25–6
vocational programmes 109
voluntariness (access) 141

Wals, A. 121
water purification technique 61–2
water scarcity 19
water vapour 18
weather, extreme 18, 23, 25, 157, 172
Wesselink, A. 183
'West' (concept) 16
Western science, mainstream 44, 45, 128
Western knowledge/traditions 117–18, 122, 150
wind energy 40, 132, 194
'woke' generation 110
workplaces 65, 87, 92, 134, 181
'work-ready' graduates 87, 92, 134
World Bank 84
Nurturing Excellence in Higher Education Program (Nepal) 85
'world-class university' 82, 87, 195–6